Biotech's

DICTIONARY *Of* Irrigation

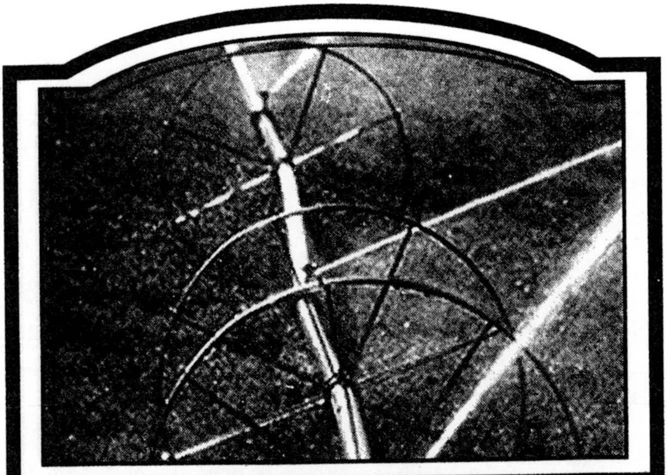

Compiled & Edited by

Meena Francis

BIOTECH

2025

BIOTECH BOOKS

Biotech's

DICTIONARY *Of* Irrigation

Compiled & Edited by
Meena Francis

First Indian Edition 2025

ISBN 978-81-7622-129-0

Published by : **BIOTECH BOOKS**
4762-63/23, Ansari Road, Darya Ganj,
New Delhi - 110002
Phones : 011 4300 3222
Email : biotechbooks@yahoo.co.in

PRINTED IN INDIA

PREFACE

Irrigation means artificial watering of land to sustain vegetation growth. Today it is being practised all over the world, especially in those parts where rainfall does not provide enough ground moisture, or in the areas of irregular rainfall to ensure harvest and to increase crop yields. It has helped greatly in extending the amount of arable land and the production of food throughout the world. Irrigated land produces twice the yield of non-irrigated land. There are basically four main methods of agriculture throughout the world: flood, furrow, sprinkler and drip irrigation.

India has made impressive gains in the field of agriculture. During the last thirty years, there has been an increase of more than fifty percent in the irrigated land. Besides undertaking close to seven hundred large irrigation projects on its rivers, it has tapped underground water, employing hundreds of thousands of motor driven tube wells. Irrigation has become more technical now. Most of the economically feasible opportunities of large-scale irrigation development have been exploited. Greater demands are being made on limited water resources; efficient use of available surface and groundwater supplies is becoming crucial. Earlier it was a matter with the farmer of watering the land but now it is becoming highly technical. There is a major thrust today to develop techniques to conserve land that is already under irrigation. This dictionary will help those who are dealing with technicalities related to irrigation.

Many a time excessive irrigation has caused harmful effects to crops due to excess salt in it. So a lot of care has to be taken while adopting an irrigation process.

- **acre-foot**
  the volume would cover one acre to a depth of one foot; 43,560 cubic feet, 1,233.5 cubic meters, 325,872 gallons.

- **100% coverage**
  the design goal of all sprinkler system designers. Sometimes used (incorrectly) in place of the term 'head to head coverage'. 100% coverage is the objective of head-to-head coverage. But head-to-head coverage does not always result in 100% coverage.

- **a posteriori classification**
  a classification made based upon the results of experimentation.

- **apriori classification**
  a classification made prior to experimentation.

- **abandoned water right**
  a water right which has not been put to beneficial use for generally five or more years, in which the owner of the water right states that the water right will not be used, or takes such actions that would prevent the water from being beneficially used.

- **abandoned well**
  a well which is no longer used or a well removed from service; a well whose use has been permanently discontinued or which is in a state of such disrepair that it cannot be used for its intended purpose. Generally, abandoned wells will be filled with concrete or cement grout to protect groundwater from waste and contamination.

- **abandonment**
  failure to put a water right to beneficial use for generally five or more years, in which the owner of the water right states that the water right will not be used, or takes such actions that would prevent the water from being beneficially used. Also see **abandoned water right**.

- **abandonment of a dam**
  1. in a legal sense, abandonment is most precisely described as transfer of all rights, title and interest in a dam to the current property owner.
  2. abandonment may also involve the slow but resolute

erosion of rights to a dam by non-use, physical destruction, lack of maintenance or intent of it. In this latter instance, the court holding jurisdiction can only decide the final determination of legal abandonment.

- **abatement**
reducing the degree or intensity of, or eliminating, pollution, as a water pollution abatement programme.

- **abiota**
1. those non-living factors which are present in and affect the characteristics of a given ecosystem. Pertaining to any non-biological factor or influence, such as geological or meteorological characteristics.

- **ablation**
1. the process by which ice and snow waste away as a result of melting and/or evaporation.
2. the erosive processes by which a glacier is reduced.

- **aboveground storage tank**
a non-vehicular device (including any associated piping that is made of non-earthen materials) located on or above the surface of the ground, or on or above the surface of the floor of a structure below the ground, such as mineworking basement or vault; and designed to contain an accumulation of petroleum products.

- **abscission**
the dropping of leaves from a plant. Premature abscission in certain plant species frequently results from excessive exposure to certain air contaminants.

- **absolute humidity**
the actual weight of water vapour contained in a unit volume of the atmosphere, usually expressed in grams of water per kilogram of air.

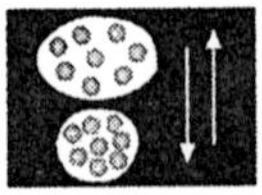

Parcel Size	Mass of H_2O Vapor	Absolute Humidity
2 m^3	10 g	5 g/m^3
1 m^3	10 g	10 g/m^3

With the same amount of water vapor in a parcel of air, an increase in volume decreases absolute humidity, while a decrease in volume increases absolute humidity.

- **absolute scarcity**
a condition when there is not enough of a resource in existence to satisfy demand for it.

- **absorb**
to take in. Many things absorb water.

- **absorber**
  a material capable of taking in a substance, such as oil, as a sponge takes up water.

- **absorption**
  1. the entrance of water into the soil or rocks by all natural processes, including the infiltration of precipitation or snowmelt, gravity flow of streams into the valley alluvium, into sinkholes or other large openings, and the movement of atmospheric moisture.
  2. the uptake of water or dissolved chemicals by a cell or an organism (as tree roots absorb dissolved nutrients in soil).
  3. more generally, the process by which substances in gaseous, liquid, or solid form dissolve or mix with other substances. Not to be confused with Adsorption.

- **absorption loss**
  the loss of water by infiltration or seepage into the soil during the process of priming, i.e., during the initial irrigation of a field, generally expressed as flow volume per unit of time.

- **abutment (of a dam)**
  the part of a valley side wall against which a dam is constructed. An artificial abutment is sometimes constructed as a concrete gravity section to take the thrust of an arch dam where there is no suitable natural abutment. Right and left abutments are designated, as one looks downstream.

- **abutment seepage**
  reservoir water that moves through seams or pores in the dam's natural abutment material and exists as seepage.

- **abyssal depth**
  in a limnological sense, that depth at which the water remains uniform in temperature, or is 'stagnant'.

- **accent plant**
  any plant which, by its placement or characteristics, differs from the surrounding landscape. It differs from a specimen plant, in that a specimen will usually standalone. Accent plants are used to break up monotony, vary the height of shrub border or direct attention to an area of the garden.

- **access**

  the way for a person to enter a lake, usually with a boat. Types of accesses include: easement access, funnel access, lake access and public access.

- **acclimatisation**

  the physiological adjustment or adaptation by an organism to new physical and/or environmental conditions. With respect to water, it is frequently used in reference to the ability of a species to tolerate changes in water temperature, degradation of water quality, or increased levels of salinity.

- **accretion**

  the slow addition to land by deposition of water-borne sediment. An increase in land along the shores of a body of water, as by alluvial deposit. Accretion and alluvion are often used synonymously.

- **ACEC**

  Area of Critical Environmental Concern.

- **acequia**

  an irrigation canal. Acequias were important forms of irrigation in the development of agriculture in the American Southwest. The proliferation of cotton, pecans and green chilli as major agricultural staples owe

  their progress to the acequia system.

- **acid**

  corrosive substances with pH of less than 7.0. Acidity is caused by high concentrations of hydrogen ions. These are

  chemicals that release hydrogen ions (H^+) in solution and produce hydronium ions (H_3O^+). Such solutions have a sour taste, neutralise bases, and conduct electricity.

- **acid aerosol**

  airborne particles composed of sulphates (SO_X), sulphuric acid (H_2SO_4), nitrates (NO_X), and/or nitric acid (HNO_3).

Dry particle diameters are typically less than 1-2 microns. Also see **acid deposition** and **acid fog**.

- **acid deposition**
  the introduction of acidic material to the ground or to surface waters. Involves a complex chemical and atmospheric phenomenon that occurs when emissions of sulphur and nitrogen compounds and other substances are transformed by chemical processes in the atmosphere, often far from the original sources, and then become deposited on the land or surface waters in either wet or dry forms. Wet deposition (commonly referred to as acid rain or acid fog) results from precipitation as rain, snow, or fog. Dry deposition results from particle fallout or acidic gases. The accumulation of acids, either in precipitation or through dry dustfall, on the land surface.

- **acid fog**
  airborne water droplets containing sulphuric acid and/or nitric acid. Typical diameters are 3-30 microns. See **acid deposition** and **acid aerosol**.

- **acid lakes**
  lakes that have water with a pH less than 6 standard units.

- **acid mine drainage (AMD)**
  acidic water that flows into streams from abandoned mines or piles of mining waste or tailings. The acid arises from the oxidation of iron sulphide compounds in the mines by air, dissolved oxygen in the water, and chemoautotrophs, which are bacteria that can use the iron sulphide as an energy source. Iron sulphide oxidation products include sulphuric acid, the presence of which has reduced or eliminated aquatic life in many streams in mining regions.

- **acid neutralising capacity**
  a measure of the ability of water or soil to resist changes in pH.

- **acid precipitation**
  atmospheric deposition (rain, snow and dryfall) that is composed of the hydrolysed byproducts from oxidised halo-

gen, nitrogen, and sulphur substances. Also see **acid rain**.

- **acid rain**

rainfall with a pH of less than 7.0. One of the principle sources is the combining of rain (H_2O) and sulphur dioxide (SO_2), nitrous oxides (NO_x), and carbon dioxide (CO_2) emissions which are bypro-ducts of the combustion of fossil fuels. These oxides react with the water vapour to form sulphuric acid (H_2SO_4), nitric acid (HNO_3), and carbonic acids (H_2CO_3). Long-term deposition of these acids is linked to adverse effects on aquatic organisms and plant life in areas with poor neutralising (buffering) capacity.

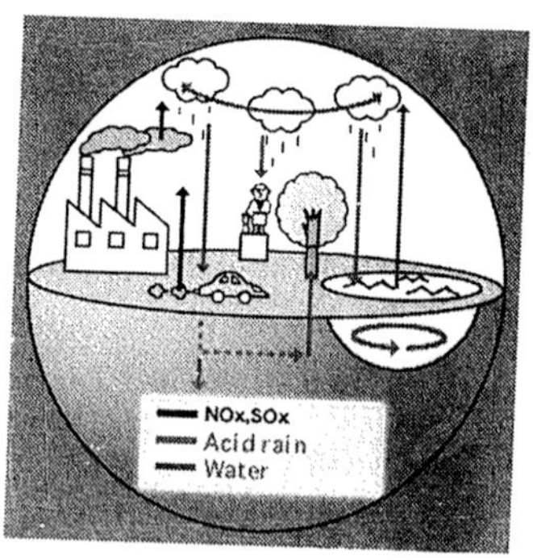

- **acid/alkaline soil**

refers to the chemical makeup of a particular soil and is measured in terms of pH. A pH of seven means the soil is neutral. A pH below seven indicates acidity and a pH above seven indicates alkalinity. Regions with light rainfall tend to have alkaline soils. Some plant species prefer a specific pH.

- **acid-forming material**

material containing sulphide minerals or other materials, which if exposed to air, water or weathering processes will form sulphuric acid that may create acid mine drainage.

- **acidic**

the condition of water or soil that contains a sufficient amount of acid substances to lower the pH below 7.0.

- **acidification**

raising the acidity (lowering the pH) of a fluid by adding an acid.

- **acidity**

a measure of how acid a solution may be. A solution with a pH of less than 7.0 is considered acidic. Solutions with a pH of less than 4.5 contain mineral acidity (due to strong inorganic acids), while a solution having a pH greater than 8.3 contains no acidity.

- **ACLs**
  Alternative Concentration Limits.

- **acre-inch**
  see **acre foot** and substitute 'inch' for 'foot'.

- **acreage**
  an area of land or water measured in acres.

- **activated carbon**
  a material produced by heating coal or wood in such a manner as to yield a porous structure, creating a very large internal surface area. Activated carbon is available in both powdered and granular forms, and is widely used to adsorb organic compounds from water and wastewater. It provides a means of removing tastes and odours from drinking water.

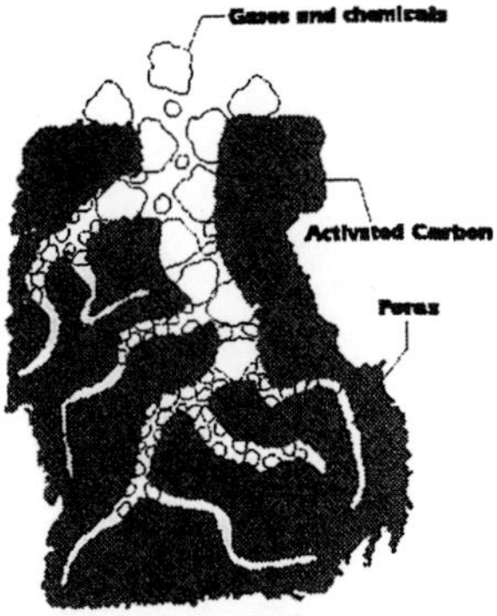

Activated Carbon adsorbs gases and chemicals

- **activated carbon adsorption**
  the process of pollutants moving out of water and attaching on to activated carbon.

- **activated sludge**
  the floc produced in raw or settled wastewater due to the growth of bacteria and other organisms in the presence of dissolved oxygen. It is the product that results when primary effluent is mixed with bacteria-laden sludge and then agitated and aerated to promote biological treatment, speeding the breakdown of organic matter in raw sewage undergoing secondary waste treatment.

- **activated sludge process**
  a method of secondary wastewater treatment, in which the waste is treated by micro-organisms in a well-aerated tank to degrade the organic material. A sedimentation tank is then used to remove the resultant sludge.

- **active solar power**
  solar energy gathered by a device that collects this energy and mechanically dis-

tributes it to where it is needed.

- **active storage capacity**
  the total amount of usable reservoir capacity available for seasonal or cyclic water storage. It is the gross reservoir capacity minus inactive storage capacity. More specifically, the volume of water in a reservoir below the maximum controllable level and above the minimum controllable level that can be released under gravity. In general, it is the volume of water between the outlet works and the spillway crest. In some instances, minimum pool operating constraints may prevent lowering the reservoir to the level of the outlet works, anf the water below the minimum pool level is not considered to be in active storage.

- **Adenosine Triphosphate (ATP)**
  an organic, phosphate-rich compound important in the transfer of energy in organisms. Its central role in living cells makes it an excellent indicator of the presence of living material in water. A measure of ATP therefore provides a sensitive and rapid estimate of biomass. ATP is reported in micrograms per litre of the original water sample.

Adenosine triphosphate (ATP)

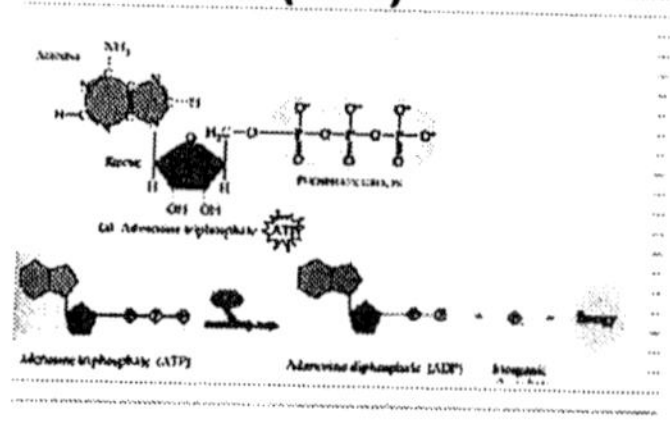

- **adequate-size farm**
  a farm with resources and productivity sufficient to generate enough income to:
  1. provide an acceptable level of family living,
  2. pay current operating expenses and interest on loans and
  3. allow for capital growth to keep pace with technological growth.

- **adhesion**
  molecular attraction that holds the surfaces of two substances in contact, such as water and rock particles. In soils, it is the process that holds water molecules

tightly to soil solids at the soil-water interfaces.

- **adit**
  a horizontal or nearly horizontal passage, driven from the surface, for the working or dewatering of a mine. Also referred to as drift, shaft, or portal.
- **adjudication**
  a judicial process whereby water rights are determined or decreed by a court of law. A court proceeding to determine all rights to the use of water on a particular stream system or ground water basin.
- **adjusted sodium adsorpvion ratio**
  index of permeability problems, based upon water quality.
- **adjusted water budget**
  quantity of water used to maintain a landscape based on evapotranspiration and area; adjusted to reflect an efficiency standard. See **water budget.**
- **administered groundwater basin**
  a groundwater basin (watershed, area, or sub-area) which, in the interest of public welfare, is monitored by an appropriate agency to ensure adequate water resources for prescribed uses. Quite often, such basins will have preferred uses designated for future development to ensure that the basin's perennial yield is not exceeded. Also referred to as designated groundwar basin.
- **adsorbate**
  any material adsorbed onto the surface of another.
- **adsorbent**
  any material which adsorbs another on its surface.
- **adsorber**
  a solid or liquid that can hold molecules of another substance on its surface.
- **adsorption**
  1. the adherence of ions or molecules in solution to the surface of solids.
  2. the adherence of a gas, liquid or dissolved material on the surface of a solid.
  3. the attraction and adhesion of a layer of ions from an aqueous solution to the solid mineral surfaces with which it is in contact. An example is the adsorption of

organic materials by activated carbon. Not to be confused with absorption.

- **advance ratio (AR)**

ratio of the time for the water to reach the end of the field to the total set time for an irrigation set on a furrow irrigation system. The ratio should be less than 0.5, in order to have good distribution uniformity.

- **advance time**

1. time required for a given stream of irrigation water to move from the upper end of a field to the lower end.
2. time required for a given surface irrigation stream to move from one point in the field to another.

- **advanced treatment**

a level of wastewater treatment more stringent than secondary treatment. It requires an 85 percent reduction in conventional pollutant concentration or a significant reduction in non-conventional pollutants.

- **advanced wastewater treatment (AWT)**

any process which reduces the level of impurities in a wastewater below that attainable through conventional secondary or biological treatment. Includes the removal of nutrients such as phosphorus and nitrogen and a high percentage of suspended solids.

- **advection**

1. the process by which solutes are transported by the bulk of flowing fluid such as the flowing ground water.
2. the horizontal transfer of heat energy by large-scale motions of the atmosphere.

- **adventive**

non-native plant that is recently introduced and is starting to spread into new places.

- **aedile**

an elected official of ancient Rome who was responsible for public works and games and who supervised markets, the grain supply and the water supply.

- **aeolian soil**

soil transported from one area to another by the wind.

- **aerate**

to supply or charge a liquid or body of water with a gas,

to expose a body of water to the circulation of air for purification. See **aerated lagoon.**

- **aerated lagoon**
  a holding and/or treatment pond that speeds up the natural process of biological decomposition of organic waste, by stimulating the growth and activity of bacteria that degrade organic waste.

- **aeration**
  any active or passive process by which intimate contact between air and liquid is assured, generally by spraying liquid in the air, bubbling air through water or mechanical agitation of the liquid to promote surface absorption of air.

- **aeration (unsaturated) zone**
  the zone between the land surface and the water table which characteristically contains liquid water under less than atmospheric pressure and water vapour and air or other gases at atmospheric pressure. The term unsaturated zone is now generally applied.

- **aeration capacity**
  volume fraction of air-filled pores in a soil at field capacity.

- **aeration tank**
  a chamber used to inject air into water.

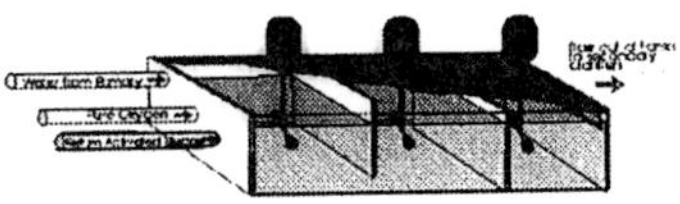

- **aerial**
  occurring above water or land.

- **aerobe**
  an organism which requires oxygen for its life processes.

- **aerobic**
  1. characterising organisms able to live only in the presence of air or free oxygen, and conditions that exist only in the presence of air or free oxygen.
  2. process requiring oxygen.

- **aerobic bacteria**
  single-celled, microscopic organisms that require oxygen to live and are partly responsible for the aerobic decomposition of organic wastes.

- **aerobic decomposition**
the biodegradation of materials by aerobic micro-organisms, resulting in the production of carbon dioxide, water and other mineral products. Generally, a faster process than anaerobic decomposition. Also see **aerobic bacteria**.

- **aerobic treatment**
the process by which microbes decompose complex organic compounds in the presence of oxygen and use the liberated energy for reproduction and growth. Such processes may include extended aeration, trickling filtration and rotating biological contactors.

- **aerodynamic**
refers to forces acting upon the soil or crop surface by moving air.

- **aeroponics**
a technique for growing plants without soil or hydroponic media. The plants are held above a system that constantly mists the roots with nutrient-laden water. Also called aero-culture.

- **aerosol**
a suspension of liquid or solid particles in air or gas.

- **aesthetics (lake)**
aesthetics are given consideration in the complete evaluation of lakes as a natural resource. The overall scenic attraction of the lake setting; natural beauty of shores and waters, or any unusual natural phenomena; the appeal of its wildlife and aquatic plants; desirable natural landscape for home sites on the shores are some of the matters considered under this heading.

- **aestival ponds**
those ponds existing only in summer.

- **AF**
acrefeet (or acrefoot).

- **affluent**
1. (lake) a tributary or feeder stream. Streams receiving the run-off from the watershed and flowing into the lake are its affluents, analogous to the affluent of a river. The analogy can be very close, where a lake has large inflowing and outflowing

streams and is located in a valley or elongated basin. See **influent** and **inlet**.
2. a stream or river that flows into a larger one; a tributary.

- **afforestation**
the artificial establishment of forest crops by planting or sowing on land that has not previously, or recently, grown trees.

- **afloat**
floating on water.

- **afterbay**
the tail race or reservoir of a hydroelectric power plant at the outlet of the turbines, used to regulate the flow below the plant. It may also refer to a short stretch of stream or conduit, or to a pond or reservoir.

- **AFY**
acre feet per year.

- **age (of groundwater)**
an approximation of the time between the water's penetration of the land surface at one location and its later presence at another location.

- **agglomeration (water quality)**
the grouping of small suspended particles into larger particles that are more easily removed through filtration, skimming or settling. Also see **coagulation**

- **aggradation**
1. the build-up of sediments at the headwaters of a lake or reservoir or at a point where streamflow slows to the point that it will drop a part or its entire sediment load.
2. modification of the earth's surface in the direction of uniformity of grade or slope, by deposition, as in a riverbed.

- **aggrading**
the building up of a stream channel which is flowing too slowly to carry its sediment load.

- **aggregate**
group of primary soil particles that cohere to each other more strongly than to other surrounding particles. Groups of individual soil particles held together naturally and consisting of particles of sand, silt and clay separated from each other by pores, cracks or planes of weakness. The term, soil structure, refers to this ar-

rangement of the soil in natural aggregrates. Various types of soil structure are recognised (massive, platy, prismatic, blocky, granular).

- **aggressive water**
  water which is soft and acidic and can corrode plumbing, piping and appliances.
- **agitated pit**
  a reservoir, pit, or pond that ordinarily is not stirred or aerated, but which is mixed just before emptying to suspend any settled solids.
- **agitator/mixer (water quality)**
  blades or paddles that slowly rotate in a tank to facilitate the mixing of suspended material.
- **AGNPS**
  agricultural nonpoint source pollution model.
- **AGP**
  algae growth potential.
- **agribusiness**
  the sum of all operations involved in the production, storage, processing and wholesale marketing of agricultural products.
- **agricultural**
  having to do with farming or farms.

- **agricultural capability**
  determines, given the ideal state, what a given area of land is capable of producing in terms of agricultural production and output.
- **agricultural drainage**
  1. the process of directing excess water away from the root zones of plants by natural or artificial means, such as by using a system of pipes and drains placed below ground surface level. Also referred to as subsurface drainage.
  2. the water drained away from irrigated farmland.
- **agricultural economics**
  the application of economic principles to the agribusi-ness sector of the economy.

- **agricultural land**
  land in farms regularly used for agricultural production; all land devoted to crop or livestock enterprises, for example, farmstead lands, drainage and irrigation ditches, water supply, cropland and grazing land.

- **agricultural levee**
  a levee that protects agricultural areas where the degree of protection is usually less than that of a flood control levee.

- **agricultural pollution**
  liquid and solid wastes from all types of farming, including runoff from pesticides, fertilisers and feedlots; erosion and dust from ploughing; animal manure and carcasses; and crop residues and debris.

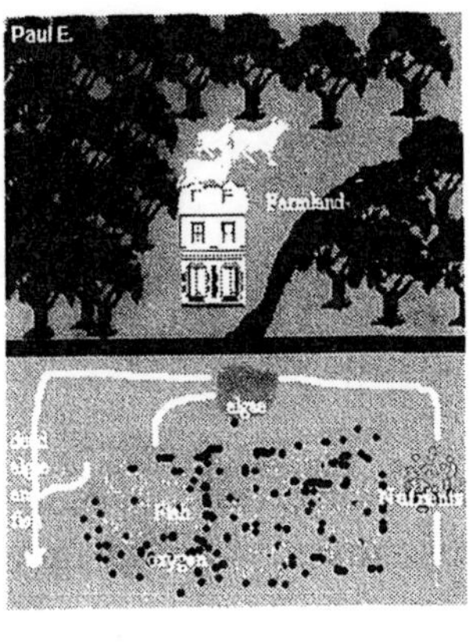

- **agricultural restructuring scenario (ARS)**
  a term used to describe the sensitivity of agricultural water demand and farm marketing revenues to changes in certain cropping patterns.

- **agricultural runoff**
  the runoff into surface waters of herbicides, fungicides, insecticides, and the nitrate and phosphate components of fertilisers and animal wastes from agricultural land and operations. It is considered as a non-point source (NPS) of water pollution. The agricultural runoff includes the water leaving areas of agricultural land use, usually enriched in nutrients, sediment and agricultural chemicals.

- **agricultural suitability**
  determines how suitable a given area of land is, in its present state, for agricultural purposes.

- **agricultural use**
  the use of any tract of land for the production of animal or vegetable life; uses include, but are not limited to, the pasturing, grazing and

watering of livestock and the cropping, cultivation and harvesting of plants.

- **agricultural water use**
  includes water used for irrigation and non-irrigation purposes. Irrigation water use includes the artificial application of water on lands to promote the growth of crops and pasture, or to maintain vegetative growth in recreational lands, parks, and golf courses. Non-irrigation water use includes water used for livestock, which includes water for stock watering, feedlots, and dairy operations, and fish farming and other farm needs.

- **agro-ecosystem**
  land used for crops, pasture, and livestock; the adjacent uncultivated land that supports other vegetation and wildlife; and the associated atmosphere, the underlying soils, ground and surface waters, irrigation channels and drainage networks.

- **agroindustrial**
  of or relating to production (as of power for industry and water for irrigation) for both industrial and agricultural purposes.

- **aground**
  onto or on a shore, reef, or the bottom of a body of water.

- **agua**
  this Spanish word for water is in names of bodies of water such as lakes and lagoons, as well as springs.

- **a-horizon**
  the uppermost zone in the soil profile, from which soluble salts and colloids are leached, and in which organic matter has accumulated. Generally, this represents the most fertile soil layer. Along with the B-Horizon, this layer constitutes part of the zone of eluviation.

- **air**
  the colourless, odourless, tasteless, gaseous mixture

that makes up the earth's atmosphere. Four gases comprise 99.997 percent (by volume) of clean, dry, air: Nitrogen (78.084 percent); Oxygen (20.946 percent); Argon (0.934 percent); and Carbon Dioxide (0.033 percent). The remaining components include neon, helium, methane, krypton, nitrous oxide, hydrogen, xenon, and various organic vapours. Under normal conditions, air contains up to about 3 percent water vapour (by volume) and many solid, liquid, or gaseous contaminants introduced by human activities and natural causes such as wind erosion and the burning of fossil fuels.

- **air binding**
  a situation where air enters the filter media and harms both the filtration and backwash processes.

- **air curtain**
  a method for mechanical containment of oils spills, in which air is bubbled through a perforated pipe, causing an upward water flow that retards the spreading of oil; also used as barriers to prevent fish from entering a polluted body of water.

- **air gap**
  physical separation of the supply pipe by at least two pipe diameters (never less than one inch) vertically above the overflow rim of the receiving vessel. In this case, line pressure is lost. Therefore, a booster pump is usually needed downstream, unless the flow of the water by gravity is sufficient for the water use. With an air gap, there is no direct connection between the supply main and the equipment. An air gap may be used to protect against a contaminant or a pollutant, and will protect against both backsiphonage and backpressure. An air gap is the only acceptable means of protecting against lethal hazards.

- **air hole**
  an opening in the frozen surface of a body of water.

- **air injection**
  in groundwater management, the pumping of com-

pressed air into the soil to move water in the unsaturated zone (Vadose Zone) down to the saturated zone (Phreatic Zone), or water table.

- **air lock**

a bubble or pocket of air or vapour, as in a pipe, that stops the normal flow of fluid through the conducting part.

- **air padding**

pumping dry air into a container to assist with the withdrawal of liquid or to force a liquefied gas such as chlorine out of the container.

- **air stripping**

a process for the removal of organic contaminants from groundwater. The groundwater flows downward inside a tower filled with materials (the packing) over a large surface area. Air is introduced at the bottom of the tower and is forced upward past the falling water. Individual organic contaminants are transferred from the water to the air, according to the gas and water equilibrium concentration values of each contaminant. Also referred to as Packed Tower Aeration.

- **air vent (of a dam)**

a pipe designed to provide air to the outlet conduit to reduce turbulence and prevent negative pressures during the release of water. Extra air is usually necessary downstream of constrictions.

- **air-bound**

condition in a pipeline wherein air trapped in a summit prevents the free flow of the material in the pipeline.

- **alfalfa valve**

outlet valve attached to the top of a pipeline riser with an opening equal in diameter to the inside diameter of the riser pipe and an adjustable lid or cover to control water flow.

- **algae**

simple single-celled (phytoplankton), colonial, or multi-celled, mostly aquatic plants, containing chlorophyll and lacking roots, stems and leaves. Aquatic algae are microscopic plants that grow in sunlit water that contains phosphates, nitrates, and other nutrients. Algae, like all aquatic plants, add oxygen to the water and are important in the fish food chain. Algae is either suspended in water (plankton) or attached to rocks and other substrates (periphyton). Their abundance, as measured by the amount of chlorophyll (green pigment) in an open water sample, is commonly used to classify the trophic status of a lake. Algae are an essential part of the lake ecosystem and provide the food base for most lake organisms, including fish. Phytoplankton populations vary widely from day to day, as life cycles are short.

- **algae blue green**
  a group of largely microscopic, photosynthetic organisms with a bacterial structure (prokaryote), but containing chlorophyll and a photosynthesis biochemistry unlike other bacteria but similar to that of other algae and higher plants. Alternative names are blue-green bacteria, cyanophytes, cyanobacteria, or, probably the most suitable, cyanoprokaryotes. Blue and red pigments, contained within them, give an often-characteristic colour.

- **algae wash**
  shoreline drift composed mainly of filamentous algae. The plants are carried to the shore by wind and wave action and stranded at or near the limit of wave advance. The drift often accumulates in considerable quantity and can become highly obnoxious on beaches.

- **algaecide**
  one of a group of plant poisons used to kill filamentous algae and phytoplankton.

- **algal bloom**
  a phenomenon whereby excessive nutrients within a river, stream or lake cause an explosion of plant life which results in the depletion of the oxygen in the

water needed by fish and other aquatic life. Algae bloom is usually the result of urban runoff (of lawn fertilisers, etc.). The potential tragedy is that of a 'fish kill', where the stream life dies in one mass extinction.

- **algal growth potential (AGP)**
  the maximum algal dry weight biomass produced in a natural water sample under laboratory conditions. Expressed as milligrams (mg), dry weight per litre (l) of sample.

- **algal growth rate**
  a measure of algal productivity in a body of water. The growth rate measures the mass of carbon used annually by algae per unit area of lake surface. The growth rate, typically referred to as primary productivity, is expressed as an index figure in grams of carbon per square metre per year, and indicates the state of eutrophication of a body of water. Algal productivity is influenced by the quantities of nutrients that flow into, or fall onto, the lake each year and the number of days of sunshine. Another important factor is the mixing of the lake, which brings up to the surface where algae exist nutrients that have accumulated near the bottom of the lake.

- **alkali**
  any strongly basic (high pH) substance capable of neutralising an acid, such as soda, potash, etc., that is soluble in water and increases the pH of a solution greater than 7.0. Also refers to soluble salts in soil, surface water, or groundwater.

- **alkali lakes**
  those containing water very highly impregnated with alkalies. The 'alkali' may be sodium carbonate or sodium sulphate and potassium carbonate but includes other alkaline compounds as well. Restricted to arid and semi-arid regions.

- **alkaline**
  sometimes water or soils contain an amount of alkali substances sufficient to raise the pH value above 7.0 and be harmful to the growth of crops. Generally, the term

alkaline is applied to water with a pH greater than 7.4.

- **alkaline (alkali) soils**
soil with pH greater than 7.0. Soil with an exchangeable sodium percentage greater than 15%. Soil that has sufficient exchangeable sodium (alkali) to interfere with plant growth and cause dispersion and swelling of clay minerals.

- **alkalinity**
the capacity of water for neutralising an acid solution. Alkalinity of natural waters is due primarily to the presence of hydroxides, bicarbonates, carbonates and occasionally borates, silicates and phosphates. It is expressed in units of milligrams per litre (mg/l) of $CaCO_3$ (calcium carbonate) or as microequivalents per litre (meq/l) 20 meq/l = 1 mg/l of $CaCO_3$. A solution having a pH below 4.5 contains no alkalinity. Low alkalinity is the main indicator of susceptibility to acid rain. Increasing alkalinity is often related to increased algal productivity. Lakes with watersheds that have sedimentary carbonate rocks are high in dissolved carbonates (hard-water lakes). Whereas lakes in granite or igneous rocks are low in dissolved carbonates (soft water lakes).

- **allelopathy**
production of substances by one organism that inhibit the growth, activity or reproduction of another.

- **allocthonous**
materials (e.g. organic matter and sediment) which enter a lake from atmosphere or drainage basin.

- **allocthonous detritus**
particulate matter originating outside, and carried into the lake.

- **allogenic**
exogenous, caused by external factors, such as a change in a habitat or environment caused by flooding.

- **allogenic succession**
predictable changes in plant and animal communities in which changes are caused by events external to the community, for example, fire, drought, floods, etc.

• **allowable depletion**
portion of plant available water that is allowed for plant use prior to irrigation based in plant and management considerations. That part of soil moisture stored in the plant root zone managed for use by plants, usually expressed as equivalent depth of water in acre inches per acre, or inches. It is sometimes referred to as allowable soil depletion or allowable soil water depletion.

• **allowable stress factor**
see **coefficient.**

• **allowable voltage loss**
voltage loss in a circuit or portion of a circuit which, if not exceeded, will result in the electrical device working correctly.

• **alluvial**
an adjective referring to soil or earth material which has been deposited by running water, as in a riverbed, flood plain or delta.

• **alluvial dam lakes**
numerous basins which are the sites of both existing and extinct lakes in the arid regions were formed by alluvial dams, especially by the coalescence of fans composed of detritus carried down by streams from opposite sides of valleys. In glaciated regions, dams were formed in valleys by glacio-fluvial deposition during the Pleistocene and barriers of various kinds, which impound water, have been created in river flood plains by alluvial deposition.

• **alluvial fan**
a fan-shaped deposit of generally coarse material created where a stream flows out onto a gentle plain. Alluvial fan is a geomorphologic feature, characterised by a cone or fan-shaped deposit of clay, silt, sand, gravel and boulders that have been eroded from mountain slopes, transported by flood flows, and deposited on the valley floor.

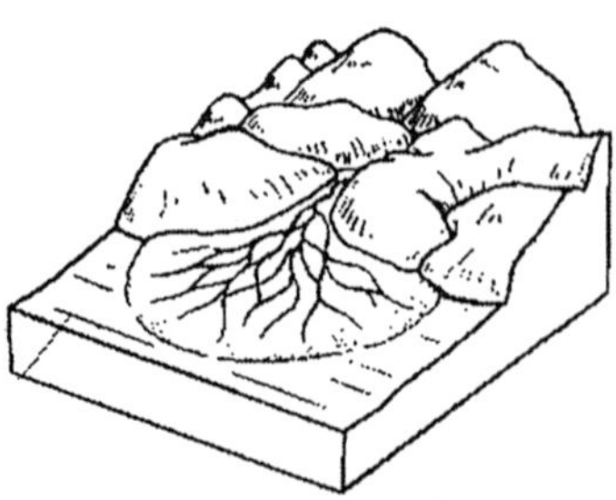

• **alluvial fan flooding**
flooding occurring on the surface of an alluvial fan or similar landform which originates at the apex and is characterised by high-velocity flows: active processes of erosion, sediment transport, deposition and unpredictable flow paths.

• **alluvial land**
areas of unconsolidated alluvium, generally stratified and varying widely in texture, recently deposited by streams, and subject to frequent flooding.

• **alluvial valley floor**
the unconsolidated stream laid deposits where water availability is sufficient for subirrigation or flood irrigation. It does not include upland areas which are generally overlain by a thin veneer of colluvial deposits composed chiefly of debris from sheet erosion, deposits by unconcentrated runoff or slopewash, talus, or other mass movement accumulation and wind-blown deposits.

• **alluvion**
1. the flow of water against a shore or bank. Inundation by water; flood.
2. the increasing of land area along a shore by deposited alluvium or by the recession of water.
in its legal meaning, alluvion is an accretion to land, made gradually, composed of detritus deposited by streams or of deposits accumulated by the action of waves and currents. On lakes, accretions are made: by alluvial deposition on the advancing front of a delta; by filling of shoreline lake bottom by erosion detritus carried by effluents, gullies and superficial surface run-off or rainwash; by shifting dune sand; and by waste disposal such as tailing from mine operations. Also, natural accretions may be made to islands, bars and beaches of lakes. The deposits are realities, but whether or not, in specific instances, they constitute legal alluvion which involves land ownership is, ultimately, a matter for court decisions.

- **alluvium**
  a general term for deposits of clay, silt, sand, gravel or other particulate material, that has been deposited by a stream or other bodies of running water in a streambed, on a flood plain, on a delta or at the base of a mountain. In lakes, alluvium is the sediments, or detrital matter carried by inflowing streams and deposited on lake bottoms. Also see **alluvion**.

- **alpine**
  the portion of mountains above tree growth or the organisms living there. Alpine vegetation is dominated by shrubs, herbs, bryophytes and lichens.

- **alpine lakes**
  lakes in any high mountain region, associated with snow, ice and a cold climate.

- **Alternate Concentration Limits (ACLs)**
  one of the three types of standards that may be applied when a leak is detected at a treatment, storage, or disposal facility and groundwater compliance monitoring is required.

- **altitude**
  altitude will affect the ability of a suction pump to pull water as the atmospheric pressure is less at higher altitudes. A rough estimate is that it will decrease about 1 foot for every 1000 feet of altitude. If you live at 20,000 feet, you will not be able to use a suction pump.

- **altocumulus cloud**
  a fleecy cloud, usually a rounded mass, but which can change radically and unexpectedly, producing intermediate forms, at an average height of 2.5 miles (4 kilometres).

- **altostratus cloud**
  a somewhat high level, blue to greyish blue cloud that forms a sheet or layer at an average height of 3.5 miles (5.6 kilometres).

- **alum treatment**
  process of putting liquid alum (Aluminium Sulphate) into the lake water, to precipitate to a floc that settles through the water column removing fine particles to the sediment and building up a barrier layer to contain soluble phosphorus in the lake sediments.

- **aluminium**
  a light, bluish white malleable and ductile metallic element found only in combinations. (Symbol Al).

- **aluminium sulphate**
  a white crystalline compound, $Al_2(SO_4)_3$, used chiefly in paper making, water purification, sanitation, and tanning.

- **alvusion**
  a sudden or perceptible change in a river's margin, such as a change in course or loss of banks due to flooding.

- **ambient water quality standards**
  the allowable amount of materials, as a concentration of pollutants, in water. The standard is set to protect against anticipated adverse effects on human health or welfare, wildlife or the environment, with a margin of safety in the case of human health.

- **AMC**
  Antecedent Moisture Condition

- **AMD**
  Acid Mine Drainage

- **amictic lake**
  a lake that does not experience mixing or turnover on a seasonal basis.

- **ammonia**
  a form of nitrogen found in organic materials, sewage and many fertilisers. It is the first form of nitrogen released when organic matter decays. It can be used by most aquatic plants and is therefore an important nutrient. It converts rapidly to nitrate (NO_3) if oxygen is present. The conversion rate is related to water temperature. Ammonia is toxic to fish at relatively low concentrations in pH-neutral or alkaline water. Under acid conditions, non-toxic ammonium ions (NH_4^+) form, but at high pH values the toxic ammonium hydroxide (NH_4OH) occurs. The water quality standard for fish and aquatic life is 0.02 mg/l of NH_4OH. At a pH of 7 and a temperature of 68 Deg. F (20 Deg. C), the ratio of ammonium ions to ammonium hydroxide is 250:1 at pH 8, the ratio is 26:1. Ammonia is

an important aquatic plant nutrient because it is readily available.

- **ammonia stripping**
  a process for the removal of ammonia from wastewater. The waste is first made alkaline to favour the NH_3 form, and then aerated so that exchange between the water and the atmosphere is encouraged. Stripping towers are often used with the waste trickling downward, as air is forced upward through the tower.

- **ammonification**
  the transformation of organic nitrogen to ammonia, generally by means of bacterial activity.

- **ammonium sulphate**
  a brownish-grey to white crystalline salt, $(NH_4)_2SO_4$, used in fertilisers and water purification.

- **amphibiotic**
  living in water during an early stage of development and on land during the adult stage.

- **amprometric titration**
  a means to measure concentrations of certain substances in water using an electric current that flows during a chemical reaction.

- **anabranch**
  a diverging branch of a river which re-enters the main stream.

- **anaerobic decomposition**
  the degradation of materials by anaerobic microorganisms living beneath the ground or in oxygen-depleted water to form reduced compounds such as methane or hydrogen sulphide. Generally a slower process than aerobic decomposition.

- **anaerobic digester**
  an airtight tank in which anaerobic micro-organisms decompose organic material and produce Biogas, mainly Methane. Sewage treatment plants often use anaerobic digesters to reduce the volume of sludge produced in Primary and Secondary Treatment, and they sometimes use the methane as a heating fuel.

- **anaerobic digestion**
  the degradation of organic matter by microorganisms

in the absence of oxygen, particularly as related to the treatment of sewage sludge. Sewage treatment plants often use anaerobic digesters to reduce the volume of sludge produced in primary and secondary treatment, and they sometimes use the resultant methane gas as a heating fuel.

- **analogue**

a continuously variable electrical signal representing a measured quantity. For example, electrical signals such as current, voltage, frequency, or phase used to represent physical quantities such as water level, flow and gate position.

- **analytical model**

a model that provides approximate or exact solutions to simplified forms of the differential equations for water movement and solute transport. Such models generally require the use of complex calculations and the use of computers.

- **ANC**

acid neutralising capacity.

- **anchor ice**

frazil ice that has collected on rocks on the stream bed. (Lake) Ice which extends down to and is attached, or frozen, to the lake bottom. Also, called Bottom Ice, Depth Ice, or Ground Ice.

- **angle valve**

refers to the water flow pattern into and out of the valve. Often used as control valves. Seldom used as isolation valves. The valve inlet is on the bottom of an angle valve and the outlet is on one side. Angle valves as a group tend to be very reliable and have lower friction losses than 'globe' valves, the other common style used for control valves.

- **anisotropy**

1. the condition of having different properties in different directions.
2. the condition under which one or more of the hydraulic properties of an aquifer vary according to the direction of the flow.

- **annual**

1. measured by the year.
2. a plant that completes its life cycle in one year or one

season. (Seed to, flowers, set seeds and dies)

- **annual flood**
  the highest peak discharge of a stream in a Water Year.

- **annual flood series**
  a list of annual floods.

- **annual low-flow**
  the lowest flow occurring each year, usually the lowest average flow for periods of perhaps 3, 7, 15, 30, 60, 120 or 180 consecutive days.

- **annual turnover**
  spring/fall turnover, mixing of the water in a lake, due to wind, annual cycle of air temperature and heating from the sun.

- **annular space**
  the space between two cylindrical objects, one of which surrounds the other, such as the space between the wall of the drilled hole and the casing, or between a permanent casing and the borehole.

- **annulus**
  for a well, the space between the pipe and the outer wall (casing) of the borehole, which may be a pipe also (the well casing).

- **annulus pressure**
  the positive pressure maintained by a fluid introduced between the well piping and the outer wall (casing) of the borehole of an underground injection well providing an indication of the integrity of the well.

- **anoxia, functional**
  although not well defined, generally refers to a body of water sufficiently deprived of oxygen so that Zooplankton and fish would not survive there.

- **anoxic**
  1. denotes the absence of oxygen, as in a body of water.
  2. of, relating to or affected with anoxia; greatly deficient in oxygen; oxygenless as with water. Water without dissolved oxygen.

- **ANSWERS**
  areal nonpoint source watershed environmental response simulation.

- **antecedent moisture**
  the degree of wetness of soil at the beginning of a runoff, determined by summation of weighted daily rainfall

amounts for a period preceding the runoff.

- **antecedent moisture condition (AMC)**
  soil moisture at the onset of a rainfall event. Dry AMC conditions mean less than 1.4 inches, average is 1.4 to 2.1 inches, and wet is greater than 2.1 inches.

- **antecedent precipitation**
  precipitation which occurred prior to a particular time over a specific area or drainage basin. Usually applied as a measure of moisture in the top layer of the soil that would affect runoff from additional rainfall.

- **antecedent precipitation index (API)**
  an index of moisture stored in a basin before a storm, calculated as a weighted summation of past daily precipitation amounts. Also see **antecedent precipitation**.

- **antecedent soil water**
  degree of wetness of a soil prior to irrigation or at the beginning of a runoff period, typically expressed as an index.

- **antecedent streams**
  antecedent streams are those in place before the rising of mountain chains. As the mountains rise, the streams cut through at the same rate and so maintain their positions.

- **antediluvian**
  1. extremely old and antiquated.
  2. occurring or belonging to the era before the Flood written about in the Bible.
  3. used sometimes today to denote a public growth and water policy based on an area's natural ability to support population growth only through existing, readily available natural resources, i.e., water.

- **anthropogenic**
  involving the impact of man on nature; induced, caused or altered by the presence and activities of man, as in water and air pollution.

- **antifluoridationist**
  one who is strongly opposed to the fluoridation of public water supplies.

• **antifreeze**
a substance, often a liquid such as ethylene glycol or alcohol, mixed with another liquid, such as water, to lower its freezing point.

• **anti-seepage collar**
a projecting collar, usually of concrete, built around the outside of a pipe, tunnel or conduit or conduit under or through an embankment dam to lengthen the seepage path along the outer surface of the conduit.

• **anti-siphon valve**
a control valve with a built-in atmospheric vacuum breaker (backflow preventer). Most commonly used in residential irrigation systems.

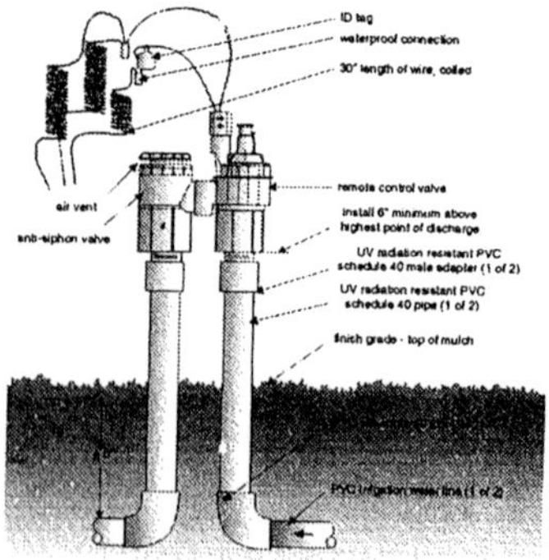

Remote control valve

• **apex**
the highest point on an alluvial fan or similar landform below which the flow path of the major stream that formed the fan becomes unpredictable and alluvial fan Flooding can occur.

• **aphotic**
1. defined as without light (Dark zone).
2. of or relating to the region of a body of water that is not reached by sunlight and in which photosynthesis is unable to occur. The Aphotic Zone of the ocean is the water deeper than about 800 meters (2,625 feet), beyond which no light penetrates. Aphotic zone is the zone in which most photosynthetic algae cannot survive, due to light deficiency. Contrast with Photic Zone.

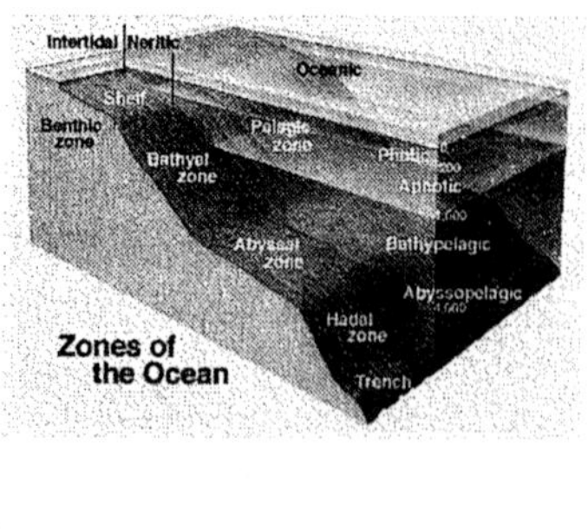

- **aphytal**
  the plantless zone of a lake bottom. See: **profundal zone**.
- **APM**
  aquatic plant management
- **apothecaries' measure**
  a system of liquid volume measure used in pharmacy.
- **application rate**
  the rate water is applied to a designated area by irrigation. Measurement is in inches-per-area for sprinklers and gallons-per-area for drip irrigation. See **water budget**.
- **application efficiency**
  the ratio of the average depth of irrigation water infiltrated and stored in the root zone to the average depth of irrigation water applied, expressed as a percent.
- **application rate**
  for irrigation, the rate at which water is applied per unit of land area, usually expressed in terms of inches per hour. See **precipitation rate**.
- **applied water**
  water delivered to a user. Also called delivered water. Applied water may be used for either inside uses or outside watering. It does not include precipitation or distribution losses. It may apply to metered or unmetered deliveries.
- **applied water demand**
  the quantity of water delivered to the intake of a city's water system or factory, the farm headgate or a marsh or other wetland, either directly or by incidental drainage.
- **appropriate**
  to authorise the use of a quantity of water to an individual requesting it.
- **appropriate technology**
  the application of current scientific knowledge and technology in such a way so as to conform with existing economic, infrastructure, social and cultural conditions and practices. By extension, the concept implies the implementation of low-technology solutions incorporating simplicity of design, use and maintenance.
- **approximate original contour**
  the surface configuration achieved by backfilling and

grading of mined areas so that the reclaimed area, including any terracing or access roads, closely resembles the general surface configuration of the land prior to strip mining and blends into and complements the drainage pattern of the surrounding terrain.

- **appurtenant structures (of a dam)**
  auxiliary features of a dam such as an outlet, spillway, powerhouse, tunnel, etc.

- **apron**
  1. a platform, as of planking, at the entrance to a dock.
  2. a covering or structure along a shoreline for protection against erosion. A platform serving a similar purpose below a dam or in a sluiceway.
  3. an area covered by sand and gravel deposited in the front of a glacial moraine.

- **aqua**
  water; an aqueous solution. A prefix meaning water, e.g., aquaculture.

- **aquaculture, also aquiculture**
  the science, art and business of cultivating marine or freshwater food fish or shellfish, such as oysters, clams, salmon and trout, under controlled conditions for commercial purposes. Examples of aquaculture plant products include rice and cranberries.

- **aquamarsh**
  a water body in which the original open water is nearly or completely obscured by emergent and floating aquatic vegetation. A stage in the evolution between open water and land marsh.

- **aquasol**
  a water soil. Water is the medium in which the plants grow.

- **aquatic**
  consisting of, relating to, or being in water; living or growing in, on, or near the water.

- **aquatic algae**
  microscopic plants that grow in sunlit water containing phosphates, nitrates and other nutrients. Algae, like all aquatic plants, add oxygen to the water and are important in the fish food chain.

• **aquatic biology**
field of biological study that deals with aquatic plants and animals.

• **aquatic ecosystem**
the basic unit of aquatic organisms and nonliving environment they live in.

• **aquatic macrophytes**
macrophytes (large plants versus microscopic) that live completely or partially in water.

• **aquatic macrophytes harvesting**
the manual and mechanical cutting and collecting of aquatic plants from lakes.

• **aquatic microbiology**
study of microscopic plants and animals and their interrelationships.

• **aquatic plant management**
steps taken to evaluate, establish a management plan, and do something to change the amount and types of aquatic plants in a lake.

• **aquatic weeds**
a common, unsightly, troublesome aquatic (water) plant, that grows in abundance or out of place.

• **aqueduct**
1. a pipe, conduit or channel designed to transport water from a remote source, usually by gravity.
2. a bridge-like structure supporting a conduit or canal passing over a river or low ground.

• **aqueous**
1. relating to, similar to, containing or dissolved in water, watery.
2. formed from matter deposited by water, as certain sedimentary rocks.

• **aqui**
a prefix for water, e.g., aquifer.

• **aquic**
a mostly reducing soil moisture regime nearly free of dissolved oxygen due to saturation by groundwater or its capillary fringe and occurring at periods when the soil tem-

perature at 50 centimetres is above 5°C (41°F).

- **aquiclude**
  an impermeable layer that confines an aquifer, preventing the water in it from moving upward or downward into adjacent strata. Shale and some igneous rocks often form aquicludes.

- **aquiclude (confining bed)**
  a formation which, although porous and capable of absorbing water slowly, will not transmit water fast enough to furnish an appreciable supply for a well or spring. Aquicludes are characterised by very low values of 'leakage' (the ratio of vertical hydraulic conductivity to thickness), so that they transmit only minor inter-aquifer flow and also have very low rates of yield from compressible storage. Therefore, they constitute boundaries of aquifer flow systems.

- **aquifer**
  a geologic formation, a group of formations, or a part of a formation that is water bearing. A geological formation or structure that stores or transmits water, or both, such as to wells and springs. Use of the term is usually restricted to those water-bearing structures capable of yielding water in sufficient quantity to constitute a usable supply.

- **aquifer compaction**
  term used to describe the effects of emptying or overdrawing an aquifer; overdrafts tend to collapse the structure of the aquifer such that the original volume cannot be restored. May also be associated with a general land subsidence in the surrounding ground level as the result of such compaction.

- **aquifer storage and recovery (ASR)**
  the injection of freshwater into a suitable aquifer during times when supply exceeds demand (wet season), and recovering it during times when there is a supply deficit (dry season).

- **aquifer system**
  a body of permeable and relatively impermeable materials that functions regionally as a water-yielding unit. It com-

prises two or more permeable units separated at least locally by confining units (aquitards) that impede ground-water movement but do not greatly affect the regional hydraulic continuity of the system. The permeable materials can include both saturated and unsaturated sections.

- **aquifer test**

  a test to determine hydrologic properties of an aquifer, involving the withdrawal of measured quantities of water from, or the addition of water to, a well and the measurement of resulting changes in head in the aquifer both during and after the period of discharge or addition (recharge).

- **aquifer, basin-fill**

  an aquifer located in a basin surrounded by mountains and composed of sediments and debris shed from those mountains. Sediments are typically sand and gravel with some clay.

- **aquifer, confined**

  an aquifer which is bounded above and below by formations of impermeable or relatively impermeable material. An aquifer in which ground water is under pressure significantly greater than atmospheric and its upper limit is the bottom of a bed of distinctly lower hydraulic conductivity than that of the aquifer itself. See **artesian aquifer**.

- **aquifer, fractured bedrock**

  an aquifer composed of solid rock, but where most water flows through cracks and fractures in the rock instead of through pore spaces. Flow through fractured rock is typically relatively fast.

- **aquifer, leaky (semi-confined)**

  an aquifer overlaid and/or underlain by a thin semipervious layer through which flow into or out of the aquifer can take place.

- **aquifer, perched**

  a groundwater unit, generally of moderate dimensions, that occurs whenever a groundwater body is separated from the main groundwater supply by a relatively impermeable stratum and by the zone of aeration above the main water body.

- **aquifer, saline/poor quality**
  an aquifer containing water that is high in total dissolved solids, and is unacceptable for use as drinking water.

- **aquifer, unconfined**
  an aquifer made up of loose material, such as sand or gravel, that has not undergone lithification (settling). In an unconfined aquifer the upper boundary is the top of the zone of saturation (water table).

- **aquifer, volcanic rock**
  an aquifer composed of rock that originated from a volcano, such as basalt. This type of rock may or may not be very permeable.

- **aquifuse**
  a formation that has no interconnected openings and hence cannot absorb or transmit water.

- **aquitard**
  a saturated, but poorly permeable bed that impedes ground-water movement and does not yield water freely to wells, but which may transmit appreciable water to or from adjacent aquifers and, where sufficiently thick, may constitute an important ground-water storage unit. Aquitards are characterised by values of leakance that may range from relatively low to relatively high. Aerial extensive aquitards of relatively low leakance may function regionally as boundaries of aquifer flow systems.

- **arable**
  having soil or topographic features suitable for cultivation.

- **arable land**
  land capable of being cultivated and suitable for the production of crops.

- **ARARS**
  applicable, relevant, appropriate requirements.

- **arboretum**
  a collection of plants, trees and shrubs grown for public exhibition, public enjoyment, recreation, education or research.

- **arboriculture**
  the planting, care and tending of trees and shrubs, individually or in small groups, for utilitarian purposes.

- **arc**
  (degrees) portion of a full circle (360°) covered by a part circle sprinkler.
- **arch dam**
  curved masonry or concrete dam, convex in shape upstream that depends on arch action for its stability; the load or water pressure is transferred by the arch to the abutments.
- **arch-gravity structure**
  a structure which derives its resistance to the pressure of water from both an arching effect and its own weight.
- **Archimedean screw**
  an ancient apparatus for raising water, consisting of either a spiral tube around an inclined axis or an inclined tube containing a tight-fitting, broad-threaded screw. Also referred to as Archimedes' screw.
- **Arctic**
  referring to the region of the earth between the North Pole and Arctic Circle.
- **Arctic tundra**
  the grassland biome characterised by permafrost (subsurface soil that remains frozen throughout the year).
- **are**
  a metric unit of land measure equal to 100 square meters or 1/100 Hectare (119.6 square yards).
- **area (sub-area), hydrographic**
  primarily these are subdrainage systems, typically valleys, within a more comprehensive drainage basin. Hydrographic areas (Valleys) may be further subdivided into hydrographic sub-areas based on unique hydrologic characteristics (e.g., differences in surface flows) within a given valley or area.
- **area flooded**
  area of a floodplain that is flooded in a specific stream reach, watershed or river basin; may be for a single flood event, but is usually expressed as an average, annual value based on the sum of areas from all individual flood events over a long period of time, such as 50 to 100 years, and adjusted to an average value.

- **area of a lake**

  the space occupied by the water surface. The area of a lake, generally, is something that cannot be determined with great exactitude; often the figure given is an arbitrary one, and figures from different sources show considerable disagreements. This comes about, because some error is inherent in any of the procedures devised for determining area; because measurements may be made from hydrographic maps which differ in accuracy and detail, and in time at which the map was made. This latter becomes important where lakes fluctuate greatly in levels. Some differences may arise also where different mathematical procedures are followed in making measurements. Also, often arbitrary decisions must be made as to location of shore line, the inclusion or exclusion of islands and boundaries between a lake and connecting water, all of which consequently affect the computed area. Area is usually expressed square miles and acres; or where the metric system is used in square kilometres and square meters.

- **area of critical environmental concern (ACEC)**

  an area on Public Lands where special management attention is required to protect and prevent irreparable damage to historic, cultural, or scenic values, fish and wildlife resources or other natural systems or processes, or to protect people from natural hazards.

- **area of influence**

  the area surrounding a pumping or recharging well within which the water table or potentiometric surface has been changed due to the well's pumping or recharge.

- **area of review**

  the area around an underground injection well that may be influenced adversely by fluid injection. Typically, the extent of this area may be calculated by using the specific gravity and rate of introduction of the injected fluids, the size, storage capacity and hydraulic conductivity of the injection zone and certain underground formation pressures.

- **area**
surface included within a set of lines. In irrigation, usually used to describe a surface of land or cross section of pipe.

- **area-capacity curve**
a graph showing the relation between the surface area of the water in a reservoir and the corresponding volume.

- **arhizous**
plant without roots.

- **arid**
1. a term applied to a climate or region where precipitation is so deficient in quantity, or occurs so infrequently, that crop production is impractical without irrigation.
2. climate characterised by less than 10 inches of annual rainfall.

- **arid climate**
see **climate.**

- **aridic**
a soil moisture regime that has no moisture available for plants for more than half the cumulative time that the soil temperature at 19.7 inches (5°0 centimetres) is above 5°C (41°F) and has no period as long as 90 consecutive days when there is moisture for plants while the soil temperature at 50 centimetres is continuously above 8°C (46.4°F).

- **aridity**
the quality or state of being arid, dry or barren.

- **arithmetic growth**
a rate of increase (or decrease) by a constant amount per time period, for example a population increase of X persons per year, year after year.

- **arithmetic mean**
the sum of a set of observations divided by the number of observations. Also referred to as simply the mean, or the sample mean.

- **arm**
1. an inlet of water (as from the sea).
2. a long and relatively narrow body of water extending inland from a main body. Usually the term arm is applied to a reach of water,

greater in length and narrower than one called a bay, but often on maps no clear distinction exists between arm and bay: nor between arm and lobe.

- **armouring**
  a facing layer (protective cover), or rip rap, consisting of very large stones placed to prevent erosion or the sloughing off of a structure or embankment. Also, a layer of large stones, broken rocks or boulders, or precast blocks placed in random fashion on the upstream slope of an embankment dam, on a reservoir shore, or on the sides of a channel as a protection against waves, ice action and flowing water. The term armouring generally refers only to very large rip rap.

- **arranged delivery**
  operation of a water delivery system to meet predetermined needs, generally based on user water orders. Also referred to as scheduled delivery.

- **arrastra**
  a crude drag-stone mill for pulverising ores, especially those containing free gold or silver; frequently powered by falling water.

- **arroyo**
  a water-carved channel or gully in an arid country which is usually rather small with steep banks and is dry much of the time due to infrequent rainfall and the shallowness of the cut, which does not penetrate below the level of permanent ground water.

- **artesian**
  a commonly used expression, generally synonymous with confined and referring to subsurface (ground) bodies of water which, due to underground drainage from higher elevations and confining layers of soil material above and below the water body (referred to as an artesian aquifer), result in underground water at pressures greater than atmospheric.

- **artesian aquifer**
  a commonly used expression, generally synonymous with (but a generally less favoured term than confined acquifer. An artesian aquifer is an aquifer which is

bounded above and below by formations of impermeable or relatively impermeable material. An aquifer in which ground water is under pressure significantly greater than atmospheric and its upper limit is the bottom of a bed of distinctly lower hydraulic conductivity than that of the aquifer itself.

- **artesian pressure**

the pressure under which artesian water in an artesian aquifer is subjected, generally significantly greater than atmospheric.

- **artesian water**

ground water that is under pressure when tapped by a well and is able to rise above the level at which it is first encountered. It may or may not flow out at ground level. The pressure in such an aquifer commonly is called artesian pressure and the formation containing artesian water is an artesian aquifer or confined aquifer.

- **artesian well**

1. a well bored down to the point, usually at great depth, at which the water pressure is so great that the water is forced out at the surface. The name is derived from the French region of Artois, where the oldest well in Europe was bored in 1126.

2. a well tapping a confined or artesian aquifer in which the static water level stands above the top of the aquifer. The term is sometimes used to include all wells tapping confined water. Wells with water levels above the unconfined water table are said to have positive artesian head (pressure) and those with water level below the unconfined water table, negative artesian head. If the water level in an artesian well stands above the land surface, the well is a flowing artesian well. If the water level in the well stands above the water table it indicates that the artesian water can and probably does discharge to the unconfined water body.

- **artesian zone**

a zone where water is confined in an aquifer under pressure so that the water rises in the well casing or drilled hole above the bottom of the confining layer overlying the aquifer.

- **articulation**
(of a lake)the ratio of area of inlets and bays to the total area of the lake.

- **artificial circulation**
the mixing of lake water using an air bubble stream or other mechanical means rather or in addition to the wind mixing.

- **artificial drains**
man-made or constructed drains.

- **artificial lakes, ponds**
basins purposely excavated by man and filled with water by catchments from runoff, by pumping or diversion of natural water bodies. Definitely artificial are those ponds constructed for farm use, for receiving factory wastes, sewage, etc. However, there are degrees of artificiality. Bodies of water impounded by dams across rivers are artificial only to a degree as are lakes whose basins have been altered by dredging or filling or whose levels has been raised or lowered respectively by dams across outlets or by dredging outlets. A gravel pit or stone quarry is patently an artificial basin, but it is filling with water, after abandonment, may be a natural process. Ponds occupying mine cave-in pits can hardly be accepted as natural geomorphological features, neither are they intentionally constructed and filled with water by man. Wherever man has made use of the water or occupied the adjacent land, he has modified natural lakes to some degree, and to that degree made them artificial.

- **artificial recharge**
the designed (by man's activities as opposed to the natural or incidental) replenishment of ground water storage from surface water sup-

plies such as irrigation or induced infiltration from streams or wells. There exist five common techniques to effect artificial recharge of a groundwater basin:

1. water spreading consisting of the basin method, stream-channel method, ditch method, and flooding method, all of which tend to divert surface water supplies to effect underground infiltration;

2. recharge pits designed to take advantage of permeable soil or rock formations;

3. recharge wells which work directly opposite of pumping wells, although they generally have limited scope and are better used for deep, confined aquifers;

4. induced recharge which results from pumping wells near surface supplies, thereby inducing higher discharge towards the well; and

5. wastewater disposal which includes the use of secondary treatment wastewater in combination with spreading techniques, recharge pits, and recharge wells to reintroduce the water into deep aquifers thereby both increasing the available groundwater supply and also further improving the quality of the wastewater.

Also referred to as Induced Recharge.

- **artificial substrate**

a device placed in the water for a specified period of time that provides living spaces for a multiplicity of organisms; for example, glass slides, concrete blocks, multi-plate samplers or rock baskets, used primarily to collect organisms in areas where the physical habitat is limiting or cannot be adequately sampled using conventional methods.

- **asexual reproduction**

plants reproducing without the sexual process by fragmentation, turions, tubers and/or other vegetative structures.

- **ASOS**

automated surface observing system.

- **aspect**

1. the compass direction toward which a sloping land area faces. The direction is measured downslope and

normal to the contours of elevation.
2. the orientation - north, south, east, west - of a plant's location. Aspect significantly influences the survival and growth behaviour of a plant. See **Micro-Climate**.

- **asperse**
to sprinkle, especially with holy water.

- **assessment report**
a comprehensive record of historical, existing and projected water quality conditions of a particular watershed.

- **assignment of water**
the transfer of a water right application or permit from one person to another. This can be done in conjunction with the sale of land.

- **assimilation**
the ability of a body of water to purify itself of pollutants.

- **assimilative capacity**
1. the ability of air, a natural body of water, or soil to effectively degrade and/or disperse chemical substances. If the rate of introduction of pollutants into the environment exceeds its assimilative capacity for these substances, then adverse effects may result to habitat and wildlife.
2. its the ability of a lake to absorb nutrients or other potential pollutants without showing averse effects.

- **athalassohaline lake**
a term used to describe a saline lake which is not of marine origin. In this respect athalassohaline lakes differ from lakes formed by the isolation of part of the ocean in as much as those lakes originating from evaporation of fresh water is the importance of bivalent ions such as calcium magnesium, and sulphate relative to the dominance of two mnovalent ions (sodium and chloride) in sea water. The different ionic ratios are important biologically because the osmotic strength of a solution at a given level of Total Dissolved Solid (TDS) changes with the valence of the dominant ions. The individual ion concentrations are also important in that some ions are more toxic to fish than others.

- **atmometer**

  an instrument used to measure the rate of evaporation.

- **atmosphere**

  the gaseous layer covering the earth. The regions of the atmosphere are the Troposphere, Stratosphere, Mesosphere, Chemosphere, and the Thermosphere (which overlaps the Ionosphere and the Exosphere). The atmosphere is one of the four components, together with the lithosphere, hydrosphere, and biosphere, that comprise the earth's ecosystem. Also see **air**.

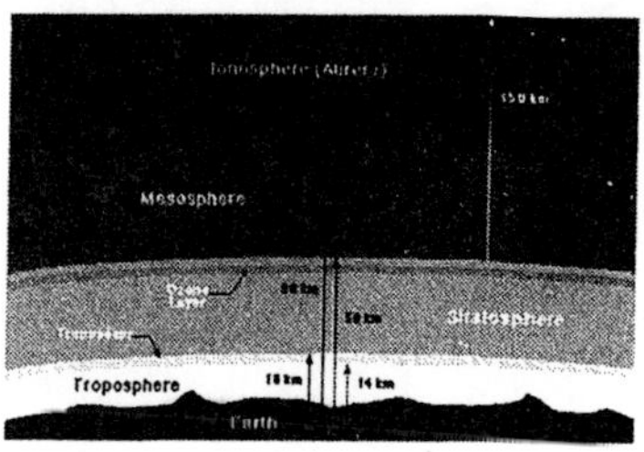

- **atmospheric pressure (PA)**

  (psi, ft water, atmospheres) absolute pressure measured at any location. Standard atmospheric pressure at sea level is defined as 14.7 psi or 34.0 ft of water.

- **atmospheric vacuum breaker(AVB)**

  backflow device configured with a single moving part, a float, which moves up or down to allow atmospheric air into the piping system. The AVB is always placed downstream from all shut-off valves. Its air inlet valve closes when the water flows in the normal direction. But, as water ceases to flow the air inlet valve opens, thus interrupting the possible backsiphonage effect. If piping or a hose is attached to this assembly and run to a point of higher elevation, the backpressure will keep the air inlet valve closed because of the pressure created by the elevation of water. Hence, it would not provide the intended protection. Therefore, this type of assembly must always be installed at least six inches above all downstream piping and outlets. Additionally, this assembly may not have shut-off valves or obstructions downstream. A shut-off valve would keep the assembly under pressure and allow the air inlet valve (or float check) to seal against the air inlet port, thus causing the assembly to act as an elbow, not a backflow preventer. The AVB may not be under continuous pressure for this same reason. An AVB must not be used for more than twelve out of any twenty-four hour period. It may be used to protect against either a pollutant or a contaminant, but may only be used to protect against a backsiphonage condition.

- **ATP**
  adenosine triphosphate.

- **Atterberg limits**
  the transition points between various states of soil consistency. The Atterberg Limits consist of:
  1. the liquid limit (water content at which the soil passes from the liquid to the plastic state)
  2. the plastic limit (water content at which the soil passes from the plastic to the semi-solid state) and
  3. the shrinkage limit (water content at which the soil passes from the semi-solid to the solid state).

- **augur mining**
  a coal-mining technique using a screw that extracts coal as it is drilled into a deposit.

- **austral**
  southern as in southern pine forest. See **boreal forest**

- **autochthonous**
  pertaining to substances (organic matter from plankton), materials or organisms originating within a particular waterway or lake and remaining in that waterway.

- **automatic valve**
  a valve which can be remotely operated. The remote operation method may be either electrical (the most common) or hydraulic. Automatic valves are commonly used as 'control valves' for irrigation systems. See **valves.**

- **autotrophic**
  plants that produce their own nutrients. If photosynthetic and green then it contains chlorophyll.
- **autumn turnover**
  the mixing of the entire water mass of a lake in the autumn.
- **autumnal**
  appearing or flowering in the fall.
- **auxiliary spillway**
  a dam spillway built to carry runoff in excess of that carried by the principal spillway; a secondary spillway designed to operate only during exceptionally large floods. Also referred to as emergency spillway. Also see **spillway**.
- **available water**
  portion of water in a soil that can be readily absorbed by plant roots. It is the amount of water released between saturated field capacity and the permanent wilting point.
- **available capacity**
  the amount of water held in the soil that is available to the plants.
- **available soil moisture (ASM) (in.)**
  difference at any given time between the actual soil moisture content in the root zone soil and the wilting point.
- **available soil moisture capacity (AMC)**
  see **available water**.
- **available water holding capacity**
  the capacity of a soil to hold water in a form available to plants. Also, the amount of moisture held in the soil between field capacity, or about one-third atmosphere of tension, and the wilting coefficient, or about 15 atmospheres of tension.
- **average annual flood damages**
  the weighted average of all flood damages that would be expected to occur yearly under specified economic conditions and development. Such damages are computed on the basis of the expectancy in any one-year of the amounts of damage that would result from floods throughout the full range of potential magnitude.

- **average annual precipitation**
  (in.)long-term historic (generally 30 years or more) arithmetic mean of precipitation (rain, snow, dew) received by an area.
- **average annual recharge**
  the amount of water entering an aquifer on an average annual basis. In many, if not most, hydrologic conditions, 'average' has little significance for planning purposes as there may exist so few 'average' years in fact.
- **average annual runoff (yield)**
  the average of water-year (October 1-September 30) runoff or the supply of water produced by a given stream or water development project for a total period of record, measured in cubic feet per second or acre-feet.
- **average discharge**
  reports on surface-water supply, the arithmetic average of all-complete water years of record whether or not they are consecutive. Average discharge is not published for less than 5 years of record. The term 'average' is generally reserved for average of record and 'mean' is used for averages of shorter periods, namely daily mean discharge.
- **average water year**
  a tern denoting the average annual hydrologic conditions based upon an extended or existing period of record. Because precipitation, runoff, and other hydrologic variables vary from year to year, planners typically project future scenarios based on hydrologic conditions that generally include average, wet (high-water), and drought (low-water) years.
- **average year water demand**
  the demand for water under average hydrologic conditions for a defined level of development.
- **average year water supply**
  the average annual supply of a water development system over a long period. For a dedicated natural flow, it is the long-term average natural flow for

wild and scenic rivers or it is environmental flows as required for an average year under specific agreements, water rights, court decisions, and congressional directives.

- **avigational trespass**
persons using a float plane to gain access to a private lake without permission, trespass first on the air or avigational rights of the lake owners, then by landing on the surface of the lake, they commit simple trespass.

- **ävja-gyttja**
lake bottom deposit composed largely of the remains of algae.

- **avoirdupois weight**
the system in common use in English-speaking countries for weighing all commodities except precious stones, precious metals, and drugs. In it 16 drams (dr.) make 1 ounce (oz.), 16 ounces make 1 pound (lb.). The pound contains 7,000 grains (453.59 grams) and is equal to 1.2153 pounds troy (or, 1 lb.=14.5833 troy oz.). There are two avoirdupois tons, the long ton (2,240 pounds) and the short ton (2,000 pounds)

- **avulsion**
1. the sudden movement of soil from one property to another as a result of a flood or a shift in the course of a boundary stream.
2. a forcible separation or detachment; a sudden cutting off of land by flood, currents, or change in course of a body of water; especially one separating land from one person's property and joining it to another's.

- **AW**
applied water

- **awash**
washed by the sea as level with or washed by waves. In such a position or way as to be covered with or as if with water.

- **AWT**
advanced wastewater treatment

- **axial flow**
fluid flow in the same direction as the axis of symmetry of the duct, vessel, or tank.

- **axis (of a dam)**
  the horizontal centreline of a dam in the longitudinal direction.

- **azotobacter**
  any of various rod-shaped, non-pathogenic, nitrogen-fixing bacteria of the genus azotobacter, found in soil and water.

- **back flow**
  any unwanted flow of used or non-potable water or substance from any domestic, industrial or institutional piping system into the pure, potable water distribution system. The direction of flow under these conditions is in the reverse direction from that intended by the system and normally assumed by the owner of the system.

- **back pressure**
  increase of pressure in the downstream piping system above the supply pressure at the point of consideration that would cause, or tend to cause, a reversal of the normal direction of flow.

- **back siphonage**
  reversal of flow (backflow) due to a reduction in system pressure which causes a negative or sub-atmospheric pressure to exist at a site in the water system. Technically, if one siphons a fluid out of a container or a pipeline, one causes that fluid to flow up over the rim of the container or top of the pipe and then down into a lower elevation through a piece of tubing or, in this case a piece of pipe that is part of the distribution system. In the vernacular, the unwanted fluid is 'sucked' into the potable water line. It is important to understand that it is not necessary for the system main to be under a true vacuum (i.e., zero psia) for backsiphonage to occur. All that is required is a negative difference in pressure and a piece of tubing or pipe that is completely full of fluid.

- **backfill**
  soil mix used to fill the plant hole after the plant is placed. Backfill often contains the excavated soil and necessary soil amendments. See **soil**

- **backflow preventer**
  a device that prevents contaminated water from being

sucked back into the water source should a reverse flow situation occur. In most places law on all irrigation systems requires backflow preventers.

- **backflow prevention device**
  any number of items such as a reduced pressure vacuum breaker, atmospheric vacuum breaker, anti-siphon valve, vacuum breaker, etc. can be used to prevent contaminates from going back into the potable water system. Backflow prevention devices should be used in all irrigation systems, especially those utilising fertiliser dispensers.

- **backpumping**
  the practice of pumping water that is leaving the area back into a surface water reservoir.

- **backsiphonage**
  reverse seepage of water in a distribution system.

- **backwashing**
  reversing the flow of water through a home treatment device filter or membrane to clean and remove deposits.

- **backwater**
  water backed up or retarded in its course as compared with its normal or natural condition of flow. In stream gagging, a rise in stage produced by a temporary obstruction such as ice or weeds, or by the flooding of the stream below. The difference between the observed stage and that indicated by the stage-discharge relation, is reported as backwater.

- **ball valve**
  this type of valve controls the water by means of a rotating ball with a hole through the centre of it. When the hole is aligned with the water flow the water flows freely through the valve with almost no friction loss. When the ball is rotated so that the hole is not aligned the flow is completely shut off. Ball valves are used primarily as isolation valves. They tend to be very reliable and trouble-free. Ball valves as a group tend to require more effort to turn on and off than other valves. For larger size pipes butterfly valves are

usually used rather than ball valves.

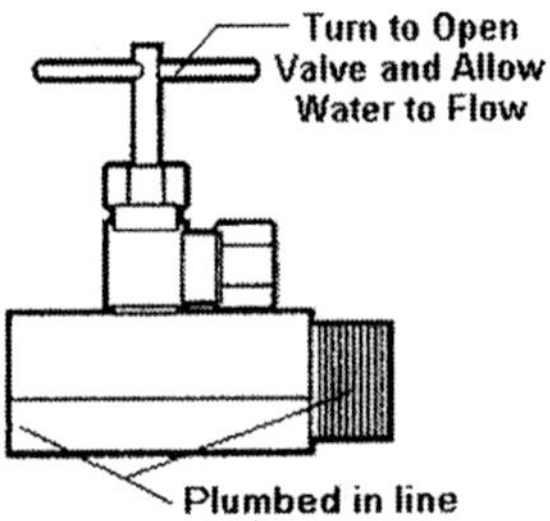

- **balled and burlapped**
  nursery stock shrubs and trees sold with a large ball of soil around the roots, wrapped in burlap to hold the rootball together, which acts as a temporary container until planted in the ground.

- **bank**
  the margins of a channel. Banks are called right or left as viewed facing in the direction of the flow.

- **bank storage**
  the water absorbed into the banks of a stream channel, when the stages rise above the water table in the bank formations, then returns to the channel as effluent seepage when the stages fall below the water table.

- **bankfull stage**
  stage at which a stream first overflows its natural banks. See also **flood stage.** Bankfull stage is a hydraulic term, whereas flood stage implies damage.

- **barrage**
  any artificial obstruction placed in water to increase water level or divert it. Usually the idea is to control peak flow for later release.

- **barrier plant**
  a plant, which by its characteristics would act as a barrier to pedestrian movement. Plants in this group generally have thorns, a dense growth pattern, and/or can tolerate a planting treatment in which they would be planted very densely.

- **base**
  a substance that has a pH of more than 7, which is neutral. A base has less free hydrogen ions (H^+) than hydroxyl ions (OH^-).

- **base discharge (for peak discharge).**
  in the Geological Survey's annual reports on surface-

water supply, the discharge above which peak discharge data are published. The base discharge at each station is selected so that an average of about three peaks a year will be presented. (See also **partial-duration flood series.**)

- **base flow**
streamflow coming from ground-water seepage into a stream. *See* **Base runoff**

- **base runoff**
sustained or fair weather runoff. In most streams, base runoff is composed largely of groundwater effluent. The term base flow is often used in the same sense as base runoff. However, the distinction is the same as that between streamflow and runoff. When the concept in the terms base flow and base runoff is that of the natural flow in a stream, base runoff is the logical term.

- **basic hydrologic data**
includes inventories of features of land and water that vary only from place to place (topographic and geologic maps are examples), and records of processes that vary with both place and time. (Records of precipitation, streamflow, ground water and quality-of-water analyses are examples.) Basic hydrologic information is a broader term that includes surveys of the water resources of particular areas and a study of their physical and related economic processes, interrelations and mechanisms.

- **basic intake rate**
see **intake rate**

- **basin**
a groundwater reservoir defined by the overlying land surface and underlying aquifers that contain water stored in the reservoir .A hydrologic unit containing one large aquifer or several connecting and interconnecting aquifers. A tract of land drained by a surface water body or its tributaries.

- **bedrock**
the solid rock beneath the soil and superficial rock. A general term for solid rock

that lies beneath soil, loose sediments, or other unconsolidated material.

- **beneficial rainfall**
  see **effective precipitation**.

- **beneficial use of water**
  the use of water for any beneficial purpose. Such uses include domestic use, irrigation, recreation, fish and wildlife, fire protection, navigation, power, industrial use, etc. The benefit varies from one location to another and by custom. What constitutes beneficial use is often defined by statute or court decisions.

- **berm**
  mounded earth forms useful in screening views and providing a sense of enclosure.

- **best management practice (BMP)**
  an irrigation BMP is a voluntary irrigation practice that is both economical and practical and is designed to reduce water consumption and protect water quality while maintaining a healthy, functional landscape.

- **biennial**
  any plant that completes its life cycle in two growing seasons. The plant usually blooms, sets seed and dies in the second growing season.

- **bioaccumulation**
  uptake and retention of substances by an organism from its surrounding medium (usually water) and from food.

- **biochemical decay**
  breakdown of pollutants in water through the action of bacteria.

- **biochemical oxygen demand (BOD)**
  the amount of oxygen used in oxidation of substances in a given water sample. Measured in milligrams per litre over a specific time period.

- **biosphere**
  the worldwide system within which all life functions, composed of smaller systems including the atmosphere, hydrosphere, and lithosphere. The earth and all its ecosystems

- **blackwater**
  wastewater from toilet, latrine, and aqua privy flushing and sinks used for food

preparation or disposal of chemical or chemical-biological ingredients.

- **Blaney-Criddle method**
  air temperature based method to estimate crop evapotranspiration.

- **blowdown**
  the water drawn from boiler systems and cold water basins of cooling towers to prevent the buildup of solids.

- **BOD**
  biochemical oxygen demand. A measure of the amount of oxygen required for neutralising organic wastes.

- **bog**
  a type of wetland that accumulates appreciable peat deposits. They depend primarily on precipitation for their water source, and are usually acidic and rich in plant matter with a conspicuous mat or living green moss.

- **bog coefficient (BC)(-)**
  inverse of scheduling coefficient but using the wettest 1% window instead of the driest. Gives no indication of the location of dry areas.

- **booster pump**
  a device to increase the water pressure is a system where some pressure already exists. For example, if water comes from a water company at 40 PSI of pressure but you need 80 PSI of pressure for the irrigation system, you would use a booster pump to increase

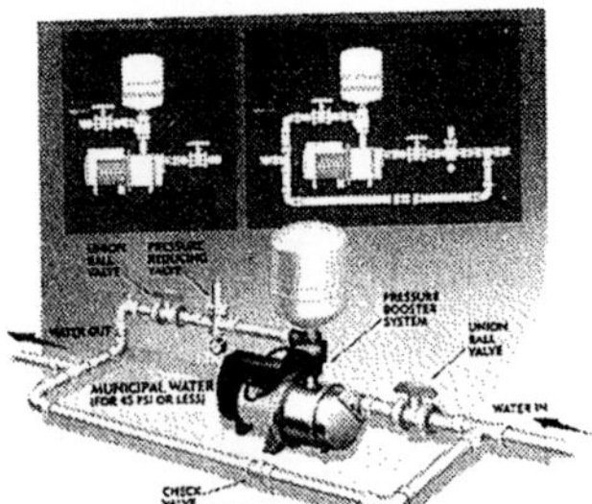

PRESSURE BOOSTER SYSTEM

the pressure. A surface pump used to increase pressure in a water line, or to pull from a storage tank and pressurise a water system.

- **border dike**
  earth ridge or small levee built to guide or hold irriga-

tion or recharge water in a field.

- **border ditch**
small excavation used as a border of an irrigated strip or plot with water being spread from one or both sides.

- **borehole**
synonym for water well.

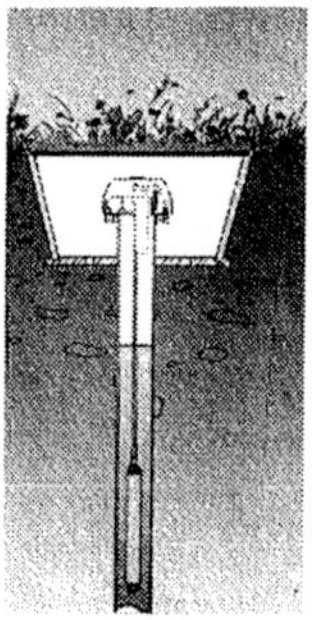

- **brackish**
1. water with a chloride level greater than 250 mg/l and less than 19,000 mg/l.
2. water containing too much salt to be useful to people but less salt than ocean water

- **braiding of river channels**
successive division and rejoining (of riverflow) with accompanying islands is the important characteristic denoted by the synonymous terms, braided or anastomosing stream. A braided stream is composed of anabranches.

- **brake horsepower hour**
see **work**.

- **brine**
highly salty and heavily mineralised water containing heavy metal and organic contaminants.

- **bubble approach**
an approach to air pollution emissions control that allows a plant to consider emissions from several sources as combined emissions from the plant.

- **bubbler**
water emission device that tends to bubble water directly to the ground or that throw water a short distance, on the order of one foot, before water contacts the ground surface.

- **bulk (density of water)**
mass of water per unit bulk volume. (approximately 1.0 g/cc, 62.4 lb/ft^3)

- **bulk density (of soil)**
mass of dry soil per unit bulk volume .(generally ranging in value from 1.3 to 1.6 g/cc)

- **buoyancy**
  the tendency of a body to float or rise when immersed in a fluid; the power of a fluid to exert an upward force on a body placed in it.

- **bushing**
  a bushing is a small piece used to connect two pipes of different sizes together. A standard reducer bushing has one male end (for the larger pipe) and one female connection (for the smaller pipe).

- **butterfly valves**
  this type of valve uses a rotating disk to control the water flow. A true butterfly valve has two half-disks, hinged together in the centre. When the disks, or 'wings' are folded together the water flows freely past them. When folded out into the water stream the wings block the flow. Most 'butterfly valves' are really 'rotating disk' valves. They have a single, round disk that rotates on an axle. When fully open the disk is rotated so that it is aligned with the water flow. To close, the disk is rotated at a right angle so that it fully blocks the flow. Butterfly valves are used as both isolation and control valves. Butterfly valves tend to be very reliable and trouble free. They are mostly used on larger pipe

Butterfly Valves

  sizes, seldom less than 3" in size. Ball valves are used on smaller size pipes.

- **c pipe**
  asbestos-cement pipe was commonly used for buried pipelines. It combines strength with lightweight and is immune to rust and corrosion.

- **cable splice**
  a joint in electrical cable. A submersible splice must be made using special materials available in kit form.

- **calcium carbonate**
  $CaCO_3$ - a white precipitate that forms in water lines,

water heaters and boilers in hard water areas; also known as scale.

- **calorie**
amount of energy required to raise the temperature of 1 gram of water 1 degree Celsius.

- **cancelled water right**
a water right that is made void either at the request of the water right holder or due to forfeiture.

- **capillarity**
the process by which water rises through rock, sediment or soil caused by the cohesion between water molecules and an adhesion between water and other materials that 'pulls' the water upward .

- **capillary action**
the means by which liquid moves through the porous spaces in a solid, such as soil, plant roots and the capillary blood vessels in our bodies due to the forces of adhesion, cohesion and surface tension. Capillary action is essential in carrying substances and nutrients from one place to another in plants and animals.

- **capillary radius**
(C_r)(ft) additional wetted radius in soil profile beyond surface wetted radius for point source emitters.

- **capillary water**
water held in the capillary or small pores of the soil, usually with soil water pressure (tension) greater than 1/3 bar. Capillary water can move in any direction.

- **capillary zone**
soil area above the water table where water can rise up slightly through the cohesive force of capillary action. See **phreatophytes**

- **carbonates**
the collective term for the natural inorganic chemical compounds related to carbon dioxide that exist in natural waterways.

- **carcinogen**
a substance that causes cancer.

- **CARL**
conservation and recreational lands.

- **carryover soil moisture**
  moisture stored in the soil within the root zone during the winter, at times when the crop is dormant, or before the crop is planted. This moisture is available to help meet water needs of the next crop to be grown.

- **casing**
  a tubular structure intended to be watertight installed in the excavated or drilled hole to maintain the well opening and, along with cementing, to confine the ground waters to their zones of origin and prevent the entrance of surface pollutants.

- **casing**
  plastic or steel tube that is permanently inserted in the well after drilling. Its size is specified according to its inside diameter.

- **catch can grid**
  containers spaced at regular intervals for collecting water for use in a water audit (sprinkler profile test).

- **catchment area.**
  see **phreatophytes**

- **cation**
  positively charged ion which during electrolysis is attracted towards the cathode. Sodium, potassium, calcium and magnesium are the most common cations in waters and soil extracts.

- **cation exchange capacity**
  the measure of the positively charged ions in a soil matrix.
  The sum of exchangeable cations (usually Ca, Ma, K, Na, Al, H) that the soil constituent or other material can adsorb at a specific pH, usually expressed in centimoles of charge per Kg of exchanger (cmol/Kg), or milli equivalents per 100 grams of soil at neutrality (pH = 7.0), meq/100g.

- **CEC**
  see **cation exchange capacity**

- **centrifugal pump**
  a pumping mechanism that spins water by means of an 'impeller'. Water is pushed out by centrifugal force. See also **multi-**

stage. Centrifugal pumps have high flow rates with low suction.

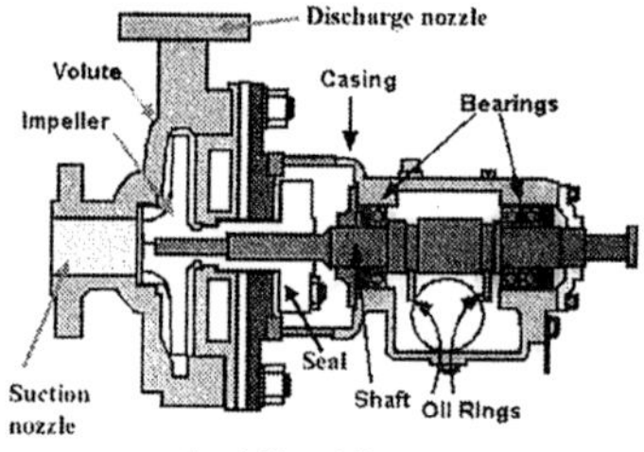

Centrifugal Pump

- **CFS**
 abbreviation of cubic feet per second.
- **CFS-day**
 the volume of water represented by a flow of 1 cubic foot per second for 24 hours. It equals 86,400 cubic feet, 1.983471 acre-feet, or 646,317 gallons.
- **CFSM (Cubic Feet per Second per Square Mile).**
 the average number of cubic feet of water per second flowing from each square mile of area drained by a stream, assuming that the runoff is distributed uniformly in time and area.
- **CFU**
 colony forming units.
- **channel**
 an open conduit either naturally or artificially created which periodically or continuously contains moving water, or which forms a connecting link between two bodies of water. River, creek, run, branch, anabranch and tributary are some of the terms used to describe natural channels. Natural channels may be single or braided *(see* **braiding of river channels**). Canal and floodway are some of the terms used to describe artificial channels.
- **channel storage**
 the volume of water at a given time in the channel or over the flood plain of the streams in a drainage basin or river reach. Channel storage is great during the progress of a flood event.
- **chaparral**
 a subtropical drought-resistant and fire-prone shrubby vegetation associated with Mediterranean-type climates.
- **check dam**
 a small dam constructed in a gully or other small water course to decrease the streamflow velocity, mini-

mise channel erosion, promote deposition of sediment and to divert water from a channel.

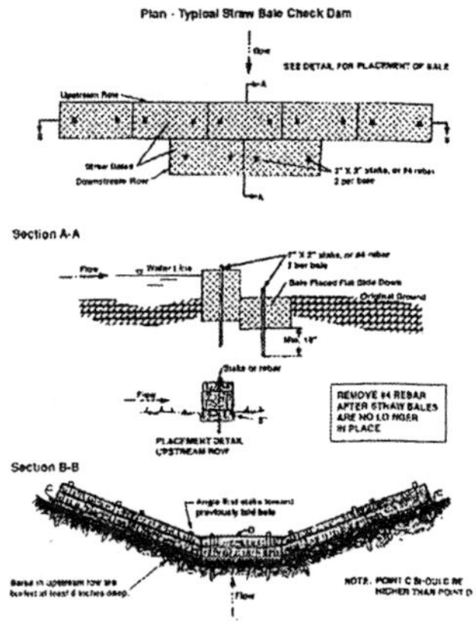

- **check valve**
  a valve that allows water to flow one way but not the other. A foot valve is one example. Electrical analogy, diode. see **valve**

- **check, check structure**
  structure to control water depth in a canal, ditch or irrigated field.

- **chemical weathering**
  attack and dissolving of parent rock by exposure to rainwater, surface water, oxygen, and other gases in the atmosphere, and compounds secreted by organisms. Contrast **physical weathering**.

- **chemigation**
  application of chemicals (including fertilisers) to crops through an irrigation system by mixing them with the irrigation water.

- **chlorination**
  the adding of chlorine to water or sewage for the purpose of disinfecting or other biological or chemical results.

- **chlorine demand**
  the difference between the amount of chlorine added to water, sewage or industrial wastes and the amount of residual chlorine remaining at the end of a specific contact period. Compare **residual chlorine.**

- **chlorosis**
  a condition in which a leaf appears more yellow than it should, especially between the veins, while the veins remain green. Chlorosis is often caused by an iron deficiency in the soil.

Interveinal Chlorosis, Red Maple (*Acer rubrum*)

- **christiansen's uniformity coefficient:**
  see **uniformity coefficient**

- **chute spillway**
  the overall structure which allows water to drop rapidly through an open channel without causing erosion. Usually constructed near the edge of dams.

- **circular mil**
  unit of measure used to report the cross sectional area of a wire conductor.

- **circulate**
  to move in a circle, circuit or orbit; to flow without obstruction, to follow a course that returns to the starting point.

- **cistern**
  a tank used to collect rainwater runoff from the roof of a house or building.

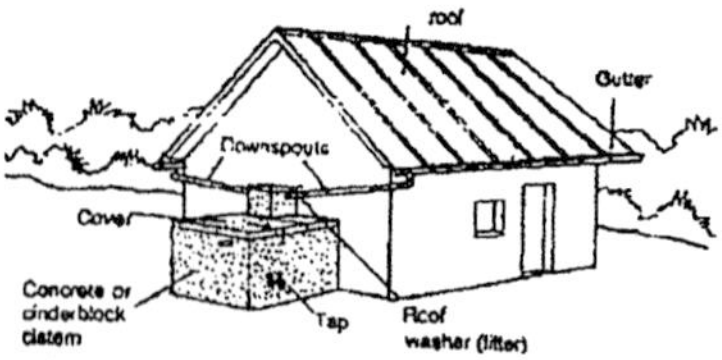

- **CL 125, 160, 200, 315**
  pronounced 'class-one-twenty-five', 'class-one-sixty', 'class-two-hundred' and 'class-three-fifteen'. Pipe classifications based on standard dimension ratios. Simply, CL 125 pipe is rated for 125 psi working pressure, CL 200 for 200 psi, etc. Most industry professionals will tell you that it's not wise to use pipe rated at less than twice the actual maximum water pressure level. In other words, for a water pressure of 100 psi use at least CL 200 pipe.

- **class (pipe)**
  term generally used to describe the pressure rating of SDR-PR (standard dimension ratio-pressure rated) PVC pipe. For example, a class 200 pipe has a pressure rating of 200 psi. (colloquial) Term used to identify the physical characteristics of thermoplastic pipe.

- **class, soil**
  group of soils defined as having a specific range in one or more particular properties such as acidity, degree of slope, texture, structure, land- use capability, degree of erosion or drainage.

- **clay**
  soil separate consisting of particles less than 0.002 mm

in equivalent diameter. Naturally occurring material composed primarily of fine-grained minerals, which is generally plastic at appropriate water contents and will harden when dried or fired.

- **clay loam**
  soil **textural class.**

- **clay soil**
  refers to the atomic arrangement of minerals that form the structure of a soil. Clay soils are composed of extremely fine minerals, usually silicates of aluminium and/or iron and magnesium.

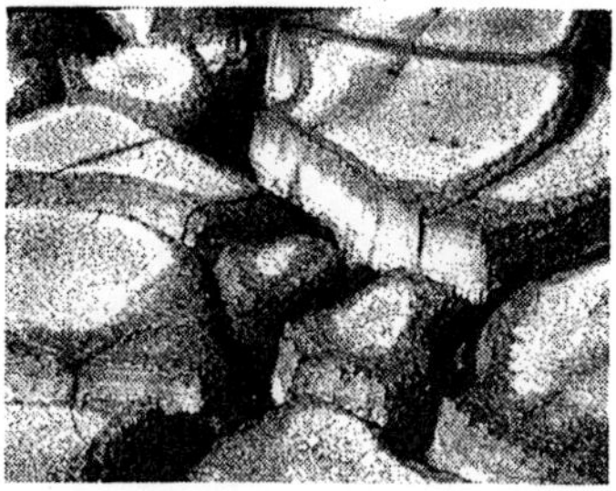

- **climate**
  the sum total of the meteorological elements that characterises the average and extreme condition of the atmosphere over a long period of time at any one place or region of the earth's surface. The collective state of the atmosphere at a given place or over a given area within a specified period of time. Generalised weather at a given place on earth over a fairly long period; a long-term average of weather.

  humid climate. Climate characterised by high rainfall and low evaporation potential. A region generally is considered as humid when precipitation averages more than 40 inches per year.

  semiarid climate. Climate characterised as neither entirely arid nor humid, but intermediate between the two conditions. A region is usually considered as semiarid when precipitation averages between 10 and 20 inches per year.

  sub-humid climate. Climate characterised by moderate rainfall and moderate to high evaporation potential. A region is usually considered subhumid when precipitation averages more than 20 inches per year, but less than 40 inches per year.

- **climatic cycle**
  the periodic changes climate displays, such as a series of dry years following a series of years with heavy rainfall.

- **climatic year**
  a continuous 12-month period during which a complete annual cycle occurs, arbitrarily selected for the presentation of data relative to hydrologic or meteorological phenomena. The climatic year is usually designated by the calendar year during which most of the 12 months occur. See **Water year**.

- **cloudburst**
  a torrential downpour of rain, which by its spottiness and relatively high intensity suggests the bursting and discharge of water from a cloud all at once.

- **coagulation**
  in water treatment, the use of chemicals to make suspended solids gather or group together into small flocs.

- **coarse sand**
  coarse sand is a soil separate. Its physical size varies according to the classification system being utilised.

- **coarse sandy loam**
  soil textural class.

- **coefficient of retardation* (c, f) (-)**
  Values describing the hydraulic frictional characteristics of a pipe material.

- **coefficient of runoff (c)(-)**
  runoff coefficient used in the rational method of predicting a design peak runoff rate. It helps to characterise runoff rate from a watershed area.

- **coefficient of uniformity**
  1. attraction of water molecules to each other.
  2. bonding strength between soil particles.
  3. attraction of like substances to one another. In soils, it is the process that forms a film of water around soil solids.

- **coefficient, consumptive use (ICUC)(%)**
  ratio of volume of irrigation water consumptively used to the total volume of irrigation water that has left the region, both in a specified period of time.

- **cohesion**
  a molecular attraction by which the particles of a body are united throughout the mass whether like or unlike. Compare **adhesion**.

- **cold vapour**
  method to test water for the presence of mercury.

- **coliform bacteria**
  bacteria of the species Escherichia coli, commonly occurring in the digestive tracts of animals; used as an indicator of the potential for disease-causing organisms in water. Non-pathogenic micro-organisms used in testing water to indicate the presence of pathogenic bacteria.

- **collector well**
  a well located near a surface water supply used to lower the water table and thereby induce infiltration of surface water through the bed of the water body to the well.

- **colloids**
  finely divided solids which will not settle but which may be removed by coagulation or biochemical action.

- **combined sewer**
  a sewer system that carries both sanitary sewage and stormwater runoff. When sewers are constructed this way, wastewater treatment plants have to be sized to deal with stormwater flows and oftentimes some of the water receives little or no treatment. Compare **separate sewer**.

- **common property resource**
  resource such as air, oceans, or sunshine that is in theory owned by everyone but in practice utilised by a few. The question of regulation arises to prevent or lessen resource abuse.

- **compaction**
  see **soil compaction**

- **compensating emitter**
  designed to discharge water at a constant rate over a wide range of lateral line pressures.

- **compensating emitter**
  see **emitter**.

- **complete fertiliser**
  a plant food consisting of all three primary nutrient elements - nitrogen, phosphorus and potassium, or (NPK).

Fertiliser packaging has these three elements listed as percentages, always in the same order - (NPK).

- **completion**
  sealing off access of undesirable water to the well bore by proper casing and/or cementing procedures.

- **composite sample, weighted**
  a sample composed of two or more portions collected at specific times and added together in volumes related to the flow at time of collection. Compare **grab sample.**

- **compost**
  the end product of the decomposition process of organic materials. Used as a soil additive, not a nutrient, compost is an amendment that improves the structure of the soil.

- **concentration**
  in the context of air or water quality, the amount of a substance per unit (weight or volume) of air or water.

- **concentration time**
  see **time of concentration**

- **concordant flows**
  flows at different points in a river system that have the same recurrence interval, or the same frequency of occurrence. It is most often applied to floodflows.

- **condensation**
  the process by which water changes from the vapour state into the liquid or solid state. It is the reverse of evaporation. The change of state from a gas to a liquid. Compare **evaporation**, **sublimation**.

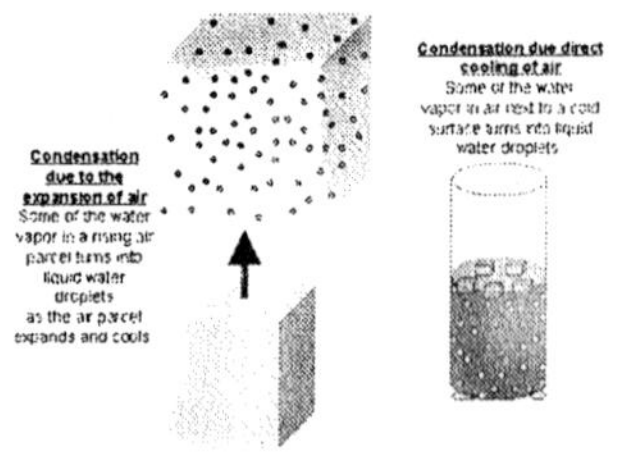

- **conduit**
  a natural or artificial channel through which fluids may be conveyed. Any open or closed channel intended for the conveyance of water.

- **cone of depression**
  a cone-shaped depression in the water table around a well or a group of wells. The cone is created by withdrawing ground water more quickly than it can be replaced. Natural depression

in the water table around a well during pumping.

- **cone of influence**
the area around a producing well which will be affected by its operation.

- **confined aquifer**
an aquifer that is bound above and below by dense layers of rock and contains water under pressure.

- **confined ground water**
a body of ground water covered by material so impervious as to sever the hydraulic connection with overlying ground water except at the intake. Confined water moves in pressure conduits due to the difference in head between intake and discharge areas of the confined water body.

- **confining bed or unit**
a body of impermeable or distinctly less permeable material stratigraphically adjacent to one or more aquifers.

- **confluent growth**
in coliform testing, abundant or overflowing bacterial growth which makes accurate measurement difficult or impossible.

- **conifer**
plants bearing their seeds in cones or modified cone-like structures. These plants are

often called evergreens, although this is not always the case. Examples would be larch, juniper and pine.

- **conjunctive management**
integrated management and use of two or more water resources, such as an aquifer and a surface water body.

- **conjunctive use**
the coordinated use of surface water and groundwater resources.

- **connate growth**
water trapped in the pore spaces of a sedimentary rock at the time it was de-

posited. It is usually highly mineralised.

- **conservation**
  the wise use or careful management of resources to attain the maximum possible social benefits from them. To protect from loss and waste. Increasing the efficiency of energy use, water use, production or distribution.

- **conservation rate structure**
  a water rate structure that is designed to conserve water. Examples of conservation rate structures include but are not limited to, increasing block rates, seasonal rates and quantity-based surcharges.

- **conservation storage**
  storage of water for later release for useful purposes such as municipal water supply, power or irrigation in contrast with storage capacity used for flood control.

- **conservation tillage**
  an agricultural system using tillage techniques designed to reduce soil erosion and overland flow. Most conservation tillage techniques involve less manipulation of the soil than conventional techniques, leaving more plant matter on the soil surface.

- **consolidated formation**
  naturally occurring geologic formations that have been lithified (turned to stone). The term is sometimes used interchangeably with the term 'bedrock'. Commonly, these formations will stand at the edges of a borehole without caving.

- **construction**
  the pan/tub has two 0.5 to 0.625 inch diameter holes drilled 3.5 inches from the top of the tub and on opposite sides. The pan is levelled and then filled until water flows from the exit holes on the side of the tub. The backboard with the mounting unit is then attached to the pan.
  The screen material selected for use is based on the crop to be grown and imposes a necessary limitation on the rate of surface evaporation. Screen materials have been

tested between 2-inch diameter mesh (chicken wire) and 1/16-inch diameter mesh (window screen). Basically, the smaller the hole is the greater is the limitation on evaporation. The screen also helps prevent animals from getting to the water surface. Preliminary research has indicated that 2-inch diameter mesh opening screen (chicken wire) is sufficient for peanut. For cotton, a combination of 2-inch screen and window screen is recommended.

- **consumer theory of value**
  an approach to valuing commodities based on how much a consumer is willing to pay for them.
- **consumptive use**
  the quantity of water absorbed by the crop and transpired or used directly in the building of plant tissue together with that evaporated from the cropped area.
- **consumptive use, net**
  the consumptive use decreased by the estimated contribution by rainfall toward the production of irrigated crops. Net consumptive use is sometimes called crop irrigation requirement.
- **consumptive waste**
  the water that returns to the atmosphere without benefiting man.
- **contact recreation**
  activities involving a significant risk of ingestion of water, such as wading by children, swimming, water skiing, diving and surfing. Compare noncontact recreation.
- **contamination**
  the introduction into water of sewage or other foreign matter that will render the water unfit for its intended use.
- **contents**
  the volume of water in a reservoir. Unless otherwise indicated reservoir content is computed on the basis of a level pool and does not include bank storage.
- **contingent valuation method**
  a method for determining the value of a resource by asking how much they would be willing to pay for it under certain circumstances.

- **continuous flushing emitter**
designed to continuously permit passage of large solid particles while operating at a trickle or drip flow thus reducing filter fineness requirements.

- **continuous-flow irrigation**
system of irrigation water delivery where each irrigator receives his allotted quantity of water at a continuous rate.

- **contour ditch**
irrigation ditch laid out approximately on the contour.

- **contour farming**
system of farming used for erosion control and moisture conservation whereby field operations are performed approximately on the contour.

- **contour flooding**
method of irrigation by flooding from contour ditches.

- **contour ploughing**
a soil conservation technique involving ploughing parallel to the contour, across a slope rather than up and down it.

- **contractor**
a water agency that signs a contract to acquire given amounts of water from another agency, usually under specified conditions.

- **control**
a natural constriction of the channel, a long reach of the channel, a stretch of rapids, or an artificial structure downstream from a gagging station that determines the stage-discharge relation at the gage.
A control may be complete or partial. A complete control exists where the stage-discharge relation at a gagging station is entirely independent of fluctuations in stage downstream from the control. A partial control exists where downstream fluctuations have some effect upon the stage-discharge relation at a gagging station. A control, either partial or complete, may also be shifting.

Most natural controls are shifting to a degree, but a shifting control exists where the stage discharge relation experiences frequent changes owing to impermanent bed or banks.

- **control structure**
  water regulating structure, usually for open conduits. A man-made structure designed to regulate the level and/or flow of water in a canal (e.g., weirs, dams).

- **controller**
  a 'timer' used to turn on and off an automatic irrigation system. Controllers range from very simple to extremely sophisticated computer systems that utilise modems, cellphones, or radios and allow 2-way communication between the controller and the units (valves, meters, weather stations, soil moisture sensors, etc.) being controlled.

- **conversion factor**
  a mathematical relation generally used in reference to water budgeting. A decimal fraction used to convert inches of evapotranspiration to gallons (0.623) or to convert inches of evapotranspiration to 100 cubic feet of water (0.00083).

- **converter**
  an electronic device for DC power that steps up voltage and steps down current proportionally (or vice-versa)..

- **conveyance loss**
  water that is lost in transit from a pipe, canal or ditch by leakage or evaporation. Generally, the water is not available for further use, however, leakage from an irrigation ditch, for example, may percolate to a ground-water source and be available for further use.

- **conveyance system efficiency**
  the ratio of the volume of water delivered to users in proportion to the volume of water introduced into the conveyance system.

- **cool season grass**
  grass that does not ordinarily lose its colour unless the average air temperature drops below 32 degrees Fahrenheit (0°C) for an extended period of time. It grows actively in

cool weather and more slowly in heat, and is not usually damaged by sub-freezing temperatures.

- **cooling tower**
  large tower used to transfer the heat in cooling water from a power or industrial plant to the atmosphere either by direct evaporation or by convection and conduction.

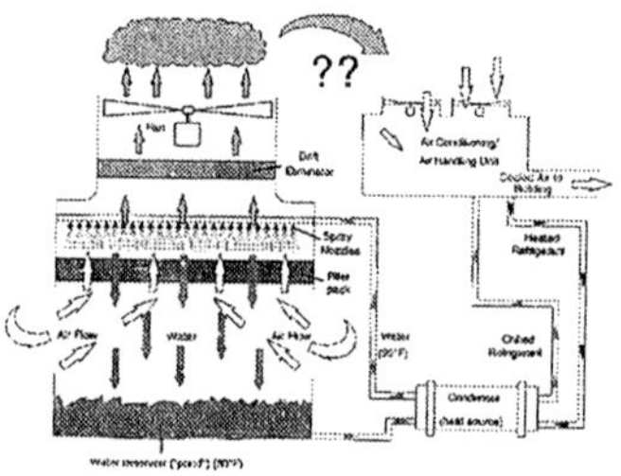

- **correlation.**
  the process of establishing a relation between a variable and one or more related variables. Correlation is simple if there is only one independent variable; multiple, if there is more than one independent variable. For gagging station records, the usual variables are the short-term gagging-station record and one or more long-term gagging-station records.

- **correlative estimate**
  a discharge determined by correlation. A correlative estimate represents a likely value of the discharge for any particular period—commonly a month—according to a specified method of analysis.

- **correlative rights**
  certain rights of land owners over a common ground water basin are coequal, or correlative, so that any one owner cannot take more than his share even if the rights of others are impaired.

- **corrugation irrigation**
  see **irrigation systems**.

- **cost theory of value**
  an approach to valuing commodities based on the cost of production.

- **cost-effectiveness analysis**
  an analysis of all the costs involved in taking a specified action to determine the most efficient way to carry out the chosen action.

- **coupler (sprinkler)**
  device, either self-sealed or mechanically sealed, that

connects the ends of two lengths of pipe or pipe to a hose.

- **coupling**
  a fitting used to join two sections of pipe together.

- **creek**
  a small stream of water which serves as the natural drainage course for a drainage basin. The term is relative according to size. Some creeks in a humid region would be called rivers if they occurred in an arid area.

- **crest**
  the top of a dam, dike, or spillway, which water must reach before passing over the structure; the summit or highest point of a wave; the highest elevation reached by flood waters flowing in a channel.

- **crisis management**
  a form of decision making that is a seat-of-the-pants response to the issue of the moment with no consideration of longer-term effects.

- **critical ground water area**
  an area that has certain ground water problems, such as declining water levels. These areas are usually limited in their development and use.

- **critical low flow**
  low flow conditions below which some standards do not apply. The impacts of permitted discharges are analysed at critical low-flow.

- **critical water supply problem areas**
  areas that have experienced, or are anticipated to experience water supply problems in the next 20 years.

- **criticality**
  a term used to describe environmental regions that are so degraded that economic activity and human habitation are impossible in the short term.

- **crop evapotranspiration**
  see **evapotranspiration.**

- **crop growth stages**
  periods of like plant function during the growing season. Usually four or more

periods are identified. They are:

initial- between planting or when growth begins and approximately 10 percent ground cover.

crop development- between about 10 percent ground cover and 70 or 80 percent ground cover.

mid season- from 70 or 80 percent ground cover to beginning of maturity.

late- from beginning of maturity to harvest.

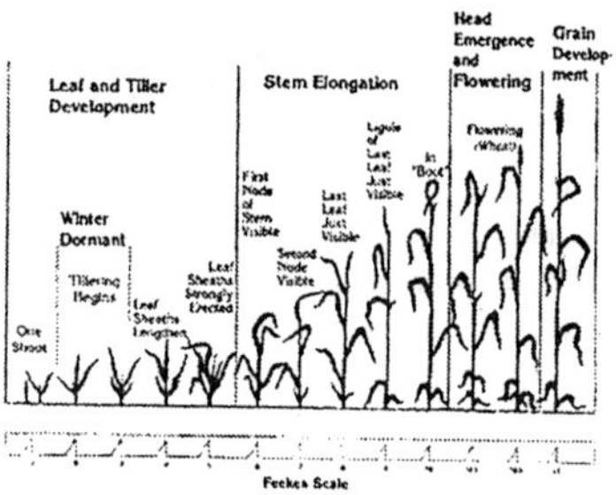

- **crop irrigation requirement**

quantity of water, exclusive of effective precipitation that is needed for crop production.

- **crop root zone**

the soil depth from which a mature crop extracts most of the water needed for evapotranspiration. The crop root zone is equal to effective rooting depth and is expressed as a depth in inches or feet. This soil depth may be considered as the rooting depth of a subsequent crop, when accounting for soil moisture storage in efficiency calculations.

- **crop rooting depth**

see **root zone**

- **crop rotation**

a soil conservation technique involving changing crops grown on a given parcel of land from year to year. Crop rotations may include fallow periods.

- **crop water requirement**

crop consumptive use plus the water required to provide the leaching requirements.

- **crop water stress index (CWSI)(-)**

index of moisture in a plant compared to a fully watered plant, measured and calculated by a CWSI instrument. Relative humidity, solar radiation, ambient air temperature and plant canopy temperature are measured.

- **crop water use**
  same as plant water requirement and evapotranspiration.

- **cropland**
  land in which crops are regularly planted and harvested. It includes land in fallow or pasture as part of a regular rotation system.

- **cropping pattern**
  the acreage distribution of different crops in any one year in a given farm area such as a county, water agency or farm. Thus, a change in a cropping pattern from one year to the next can occur by changing the relative acreage of existing crops, and/or by introducing new crops, and/or by cropping existing crops.

- **cross**
  a fitting that joins 4 sections of pipe at one point forming a 'cross'. Reducing crosses are available which have different size outlets. Unless you order a custom made cross the outlets opposite each other are always the same size.

- **cross connection**
  any actual or potential connection or structural arrangement between a public or private potable water system and any other source or system through which it is possible to introduce into any part of the potable system any used water, industrial fluids, gas or substance other than the intended potable water with which the potable system is supplied. By-pass arrangements, jumper connections, removable sections, swivel or change-over arrangements or other temporary arrangements through which backflow could occur are considered to be cross-connections. See also **backflow.**

- **crown injury**
  permanent damage to the leaf-bearing portion of a plant.

- **cryology**
  science of ice and snow.

- **cubic feet**
  a measurement of water quantity, a cubic foot is one foot in length, one foot in width and one foot deep

- **cubic feet per second**
  a unit expressing rates of discharge. One cubic foot per second is equal to the discharge of a stream of rectangular cross section, 1 foot wide and 1 foot deep, flowing water an average velocity of 1 foot per second.

- **cubic metres**
  a metric measurement of water quantity, often used by water companies to measure water use by customers. A cubic meter is one meter in length, one meter in width and one meter deep.

- **cultivar**
  a hybrid name derived from cultivated variety and applied to a plant that has originated and persisted under cultivation, as distinct from a species. It is usually distinguished by the name in quotation marks Acer platanoides 'Emerald Queen.'

- **cumulative intake**
  depth of water absorbed by soil from the time of initial water application to the specified elapsed time.

- **current**
  the portion of a stream or body of water which is moving with a velocity much greater than the average of the rest of the water. The progress of the water is principally concentrated in the current.

- **current (i)(amperes, amps)**
  movement or flow of electrons. The flow of electrons in a conductor.

- **current meter**
  an instrument for measuring the speed of flowing water.

- **cusec**
  this abbreviation for cubic foot per second.

- **cut slope**
  a landscape term describing the practice of creating level areas in the landscape by cutting into a slope and adding a retaining wall. With this process, the levelled area is generally more stable than filling in to level the slope since the soil is already compacted.

- **CV**
  abbreviation for an unknown Cultivar. See **Cultivar**.

- **cycle**
  a regularly recurring succession of events such as the cycle of the seasons. Use of cycle to describe a group of wet years followed or preceded by a group of dry years is to be avoided.

- **cycle time**
  (min, h) length of water application periods, typically used with surge irrigation.

- **dam**
  a structure of earth, rock, or concrete designed to form a basin and hold water back to make a pond, lake, or reservoir.

- **dams, locks and canals**
  providing important water transportation linkages and benefits to commerce.

- **Darcy's law**
  an equation that states that flowthrough porous media is directly proportional to hydraulic head and inversely proportional to the length of flow. Henri Darcy was a French engineer who worked at the Dijon water works in the mid 19th century. His 'law' is the basis for much of the science of ground water hydrology and one of the most important basic equations used in hydrogeologic calculations.

- **dead storage**
  the volume in a reservoir below the lowest controllable level.

- **deadhead**
  removal of spent or faded flowers to improve the plant appearance and generally ensure a longer bloom time. Typically, flowers and stems should be removed to the next lowest node.

- **deciduous**
  a woody plant that sheds its leaves before winter or drought dormancy. See **Herbaceous**.

• **decreaser**
a plant species in a range community that declines in importance as a result of grazing pressure. Usually, decreasers are the most palatable to the grazing animals.

• **deep percolation**
the movement of water by gravity downward through the soil profile beyond the root zone; this water is not used by plants.

• **deep percolation percentage (DP)(%)**
ratio of the average depth of irrigation water infiltrated and drained out of the root zone to the average depth of irrigation water applied.

• **deer antler damage**
seasonal injury to tree bark and branches caused by deer rubbing the velvet off their antlers. The rough bark edges and exposed inner layers of the trunk are more susceptible to disease and insects. Severe damage will kill the plant to below the injury.

• **deficit irrigation**
irrigation water management alternative where the soil in the plant root zone is not refilled to field capacity in all or part of the field.

• **defoliation**
refers to the unnatural loss of plant leaves. Defoliation may occur from a variety of environmental causes, such as wind, intense heat, drought, early frost, chemicals, insects or diseases.

• **deforestation**
any process of replacement of forest vegetation with other types.

• **degree of hazard**
the type of backflow preventer used to prevent backflow from occurring at the point of a cross-connection depends on the type of

substance which may flow into the potable water supply. A pollutant is considered to be any substance which would affect the colour or odour of the water, but would not pose a health hazard. This is also considered a non-health hazard. A substance is considered a health hazard if it causes illness or death if ingested. This health hazard is called a contaminant.

- **de-ionised water**
  water free of inorganic chemicals.

- **delivery box (irrigation)**
  structure diverting water from a canal to a farm unit often including measuring devices. Also called 'turn-out'. Water control structure for diverting water from a canal to a farm unit often including a measuring device. Also called delivery site, delivery facility, and turnout.

- **delivery loss**
  preferred term is conveyance loss.

- **delta**
  an alluvial deposit made of rock particles (sediment,

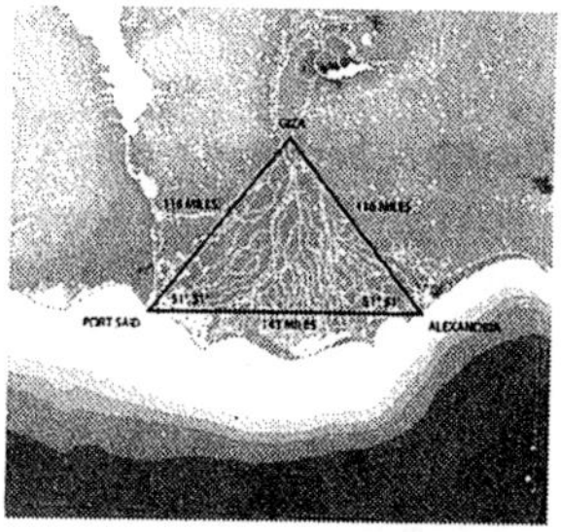

and debris) dropped by a stream as it enters a body of water.

- **demand**
  the number of units of something that will be purchased at various prices at a point in time. Compare supply. The quantity of water needed to be withdrawn to fulfil a requirement.

- **demand irrigation (system) (delivery)**
  procedure where each irrigator may request irrigation water in the amount needed and at the time desired.

- **demand management (water conservation)**
  reducing the demand for water through activities that alter water use practices, improve efficiency in water use, reduce losses of water, reduce waste of water, alter land

management practices and/or alter land uses.

- **demand scheduling**
method of irrigation scheduling whereby water is delivered to users as needed and which may vary in flow rate, frequency and duration. Considered a flexible form of scheduling.

- **demographic transition**
the process by which a human population goes through a growth pattern, including an early phase of high birth and death rates, an intermediate phase of high birth rates but low death rates, and a later phase of low birth and death rates.

- **density (of water)**
mass of water per unit volume. Same as bulk density of water.

- **densogram**
pattern of dots that shows the expected coverage from a particular combination of sprinklers, nozzles, pressure and spacing. Graphical representation of precipitation rates.

- **dental fluorosis**
disorder caused by excessive absorption of fluorine and characterised by brown staining of teeth.

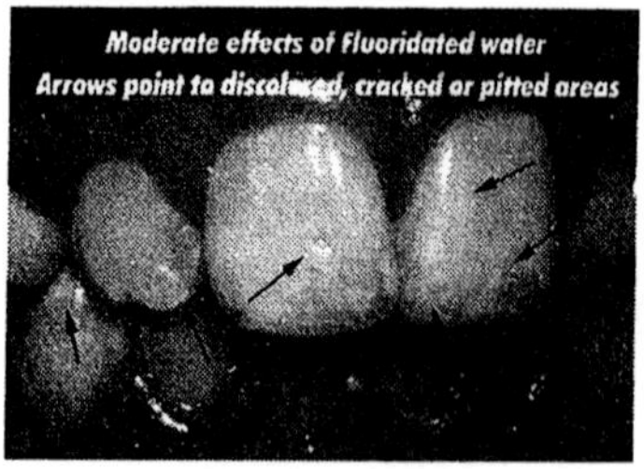

- **dependable yield, n-years**
the minimum supply of a given water development that is available on demand, with the understanding that lower yields will occur once in n years, on the average.

- **depletion**
the progressive withdrawal of water from surface or ground-water reservoirs at a rate greater than that of replenishment.

- **deposit**
something dropped or left behind by moving water, as sand or mud.

- **depression storage**
the volume of water contained in natural depressions in the land surface, such as puddles.

- **depth of irrigation**
  1. (acre inches per acre, in., ft),'depth of water applied; 2. depth of soil affected by an irrigation event.

- **depth**
  general term relating to depth of soil, water or similar item.

- **desalination**
  a process which treats saline water to remove chlorides and dissolved solids. Artificial removal of salt from water, such as by distillation or reverse osmosis.

- **desert**
  a biome characterised by plants and animals adapted to extreme moisture scarcity.

- **desertification**
  a process of land becoming more desertlike as a result of human-induced devegetation and related soil deterioration, sometimes aggravated by drought.

- **design pressure**
  in the irrigation design tutorials the design pressure is the total pressure available to operate the irrigation system. Other uses of the term vary, but usually refer to the operating pressure at which a specific piece of irrigation equipment is designed to operate.

- **detection limit**
  the lowest level that can be determined by a specific analytical procedure or test method.

- **developed water**
  water in a specific area that has been added to the hydrologic water system through engineering strategies.

- **diameter of throw**
  average diameter of the area wetted by an irrigation sprinkler operating in still air.

- **diameter**
  dimension / size of a circular pipe, usually but not always the inside diameter (ID).

inside diameter (ID, D)(in.)
outside diameter (OD, D)(in.)

- **diaphragm pump**
a type of pump in which water is drawn in and forced out of one or more chambers, by a flexible diaphragm. Check valves let water into and out of each chamber.

- **diatomaceous**
consisting of or abounding in diatoms, a class of unicellular or colonial algae having a solicified cell wall that persists as a skeleton after death.

- **dielectric union**
pipe connection (union) having an insulator between the two sides of the union for the purpose of reducing corrosion caused by galvanic action.

- **diluting water**
distilled water that has been stabilised, buffered, and aerated. Used in the BOD test.

- **dilution**
in water quality, a reduction in pollutant concentration caused by mixing with water with a lower concentration of the substance.

- **dimension ratio**
ratio of the average pipe diameter to the minimum wall thickness. The pipe diameter may be either outside or inside diameter. See related term standard dimension ratio.

- **diode**
an electronic semiconductor device that allows current flow in only one direction. Also called a rectifier. The electrical equivalent of a check valve in water.

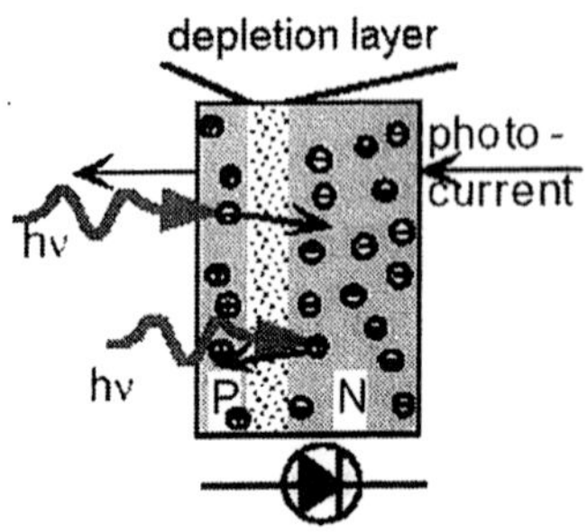

- **direct current (DC)**
current in which electrons flow constantly in one direction.

- **direct runoff**
the runoff entering stream channels promptly after rainfall or snowmelt. Superposed on base runoff, it forms the bulk of the hydrograph of a flood.

See also **surface runoff**. The terms base runoff and direct runoff are time classifications of runoff. The terms ground-water runoff and surface runoff are classifications according to source.

- **discharge**

in its simplest concept discharge means outflow; therefore, the use of this term is not restricted as to course or location, and it can be applied to describe the flow of water from a pipe or from a drainage basin. If the discharge occurs in some course or channel, it is correct to speak of the discharge of a canal or of a river. It is also correct to speak of the discharge of a canal or stream into a lake, a stream, or an ocean. the terms discharge, streamflow and runoff represent water with the solids dissolved in it and the sediment mixed with it. Of these terms, discharge is the most comprehensive. The discharge of drainage basins is distinguished as follows:

yield. Total water runout or crop; includes runoff plus underflow.

runoff. That part of water yield that appears in streams.

streamflow. The actual flow in streams, whether or not subject to regulation, or underflow.

Each of these terms can be reported in total volumes (such as acre-feet) or time rates (such as cubic feet per second or acre-feet per year). The differentiation between runoff as a volume and streamflow as a rate is not accepted.

- **discharge of ground water**

occurs when water flows out from underground.

- **disinfection**

the killing of the larger portion of the harmful and objectionable bacteria in the sewage. Usually accomplished by introduction of chlorine, but more and more facilities are using exposure to ultraviolet radiation, which renders the bacteria sterile.

- **disinfection byproducts**
  halogenated organic chemicals formed when water is disinfected.

- **dispersion**
  the movement and spreading of contaminants out and down in an aquifer.

- **dissolved oxygen (DO)**
  amount of oxygen gas dissolved in a given quantity of water at a given temperature and atmospheric pressure. It is usually expressed as a concentration in parts per million or as a percentage of saturation.

- **dissolved solids**
  substances normally solid at ambient temperatures but dissolved into ionic form in water.

- **distillation**
  water treatment method where water is boiled to steam and condensed in a separate reservoir. Contaminants with higher boiling points than water do not vaporise and remain in the boiling flask.

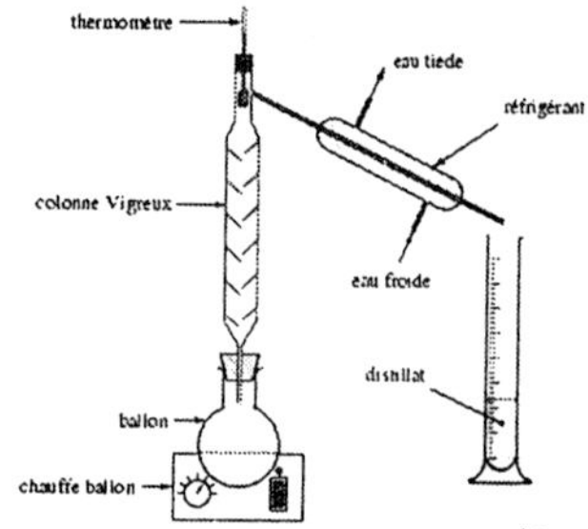

- **distilled water**
  water that has been treated by boiling and condensation to remove solids, inorganics and some organic chemicals.

- **distribution efficiency**
  measure of the uniformity of irrigation water distribution over a field.

- **distribution graph (distribution hydrograph)**
  a unit hydrograph of direct runoff modified to show the proportions of the volume of runoff that occurs during successive equal units of time.

- **distribution of water**
  the management of water which allows water users to receive the amount of water to which they are entitled by law and as supply permits.

- **distribution system**
  system of ditches or conduits and their appurtenances, which conveys irrigation water from the main canal to the farm units.

- **distribution uniformity of lower quarter**
  ratio of the average low quarter depth of irrigation water infiltrated to the average depth of irrigation water infiltrated. Ratio of the average of the lowest one-fourth of measurements of irrigation water infiltrated (or applied) to the average depth of (the total) irrigation water infiltrated (applied).

- **Distribution Uniformity (DU)**
  measure of the uniformity of irrigation water over an area.
  Measure of how uniformly water is made available to the plants over an area. DU is expressed as a percentage and generally represents the major component of irrigation efficiency. A DU of 1.0 (100% uniformity) represents all of the area infiltrating the same depth of water.

- **disturbed (manipulated) soils**
  soils with soil profiles that have been altered because of earth-moving activities (or soil amendment).

- **ditch**
  constructed open channel for conducting water.

- **diversion**
  in the most simple term it means to remove water from a water body. It can also mean individually designed diversions across a hillside. They may be used to protect bottomland from hillside runoff, divert water away from active gullies or protect buildings from runoff. Channel constructed across the slope for the purpose of intercepting surface runoff, changing the accustomed course of all or part of a stream.

- **diversion box**
  structure built into a canal or ditch for dividing the water into predetermined portions and diverting it to other canals or ditches.

- **diversion dam**
  barrier built in a stream for the purpose of diverting part

or all of the water from the stream into a canal.

- **divert**
  direct a flow away from its natural course.

- **divide**
  a ridge or high area of land that separates one drainage basin from another.

- **domestic consumption (use)**
  water used for household purposes such as washing, food preparation and showers. It is the quantity or quantity per capita, of water consumed in a municipality or district for domestic uses or purposes during a given period. It sometimes encompasses all uses, including the quantity wasted, lost or otherwise unaccounted for.

- **double check assembly (DCA)**
  two internally loaded, independently operating check valves together with tightly closing resilient seated shutoff valves upstream and downstream of the check valves. Additionally, there are resilient seated test cocks for testing of the assembly. The DCA may be used to protect against a pollutant only. However, this assembly is suitable for protection against either backsiphonage or back-pressure.

- **drainage**
  the horizontal movement of water across and through soil.

- **drainage area**
  of a stream at a specified location is that area, measured in a horizontal plane, enclosed by a topographic divide from which direct surface runoff from precipitation normally drains by gravity into the stream above the specified location.

- **drainage basin**
  the topographic land area that contributes flow to a river/ lake/wetland as defined by a particular point of reference. Sometime referred to as a catchment area. The area contributing water to a particular reference point in an aquifer system is known as a phreatic catchment. Because of geologic structure, topographic and phreatic catchment areas may not exactly coincide, especially when small areas are considered.

- **drainage density**
  length of all channels above those of a specified stream order per unit of drainage area.

- **drainage divide**
  the rim of a drainage basin. (See **Watershed**.)

- **drainage system**
  collection of surface and/or subsurface drains, together with structures and pumps, used to remove surface or groundwater.

- **drainage water**
  the water which has been collected by a drainage system. It may come from surface water or from water passing through the soil. It may be of a quality suitable for reuse or it may be of no further economic use.

- **drawdown**
  when a well is pumped, water is removed from the aquifer surrounding the well, and the water table or piezometric surface is lowered. The drawdown at a given point is the distance the water level is dropped.

- **driller's well log**
  a log kept at the time of drilling showing the depth, thickness, character of the different strata penetrated, location of water-bearing strata, depth, size and character of casing installed.

- **drip (trickle) irrigation**
  an irrigation method in which water is delivered to or near each plant in small-diameter plastic tubing. The water is then discharged at a rate less than the soil infiltration capacity through pores, perforations or small emitters on the tubing. The tubing may be laid on the soil surface, be shallowly buried, or be supported above the surface (as on grape trellises).

- **drip line**
  the imaginary circle drawn on the soil around a tree directly under its outermost branch tips. Moisture tends to drip from the tree at this point. The term is used in connection with feeding, watering and grading

around existing trees and shrubs.

- **drip system**

  an irrigation system that uses drip irrigation. See **drip irrigation** above.

- **drop pipe**

  the pipe that carries water from a pump in a well up to the surface.

- **drop spillway**

  an overfall structure in which water drops over a vertical wall onto a protected apron at a lower elevation.

- **drought**

  a period of deficient precipitation or runoff extending over an indefinite number of days, but with no set standard by which to determine the amount of deficiency needed to constitute a drought. Thus, there is no universally accepted quantitative definition of drought; generally, each investigator establishes his own definition.

  The following paragraph discusses the problem of defining a drought:

  When in an area that is ordinarily classed as humid, natural vegetation becomes desiccated or defoliates unseasonably and crops fail to mature owing to lack of precipitation, or when precipitation is insufficient to meet the needs of established human activities, drought conditions may be said to prevail. Although water for irrigation or other uses in arid areas is always limited, special shortages in such areas are also regarded as droughts. Unsatisfactory distribution of precipitation throughout the year may be as effective a factor in causing a drought as a shortage in the total amount. Temperature and wind may also play an important part, especially in relation to the damage done.

- **dry farming**

  agricultural production in climatically marginal lands without the use of irrigation.

- **dry rock geothermal energy**

  a method of extracting heat from the earth by pumping water through hot rocks.

- **dry weight (of soil sample)**

  oven dry weight of a soil sample.

- **dual purpose wells**
  wells designed with the capability of pumping water underground during artificial recharge and to the surface from the aquifer during recovery.

- **ductile iron**
  form of iron used to make pipes and fittings.

- **duplicates**
  two separate samples with separate containers taken at the same time and at the same place.

- **dynamic pressure**
  measure of water pressure with the water in motion (also known as working pressure).

- **effective porosity**
  the portion of pore space in saturated permeable material where the movement of water takes place.

- **effective precipitation**
  the part of precipitation which produces runoff, a weighted average of current and antecedent precipitation 'effective' in correlating with runoff. It is also that part of the precipitation falling on an irrigated area which is effective in meeting the requirements of consumptive use.

- **effective rainfall**
  the portion of rainfall that infiltrates the soil and is stored for plant use in the crop root zone, as calculated by the modified Blaney-Criddle model. Portion of total rainfall which becomes available for plant growth.

- **effective rainfall and irrigation**
  one of the greatest concerns in irrigation scheduling is 'how much of the water is actually going into the soil and becomes available to the plant.' That water would be termed 'effective.' Rainfall or irrigation water that does not enter the soil (runoff), or enters the soil and does not remain in the root zone where the plants can access the water (deep percolation) is considered 'ineffective'.

- **efficiency (e)(%, fraction)**
  application efficiency (E_a, AE) Ratio of the average depth of irrigation water infiltrated and stored in the root zone to the average depth of irrigation water applied.

Amount of water stored in the root zone that is available for plant use divided by the average amount of water applied during an irrigation.

- **best efficiency point (BEP)**

highest efficiency on a pump characteristic curve.

conveyance efficiency(E_c)
ratio of the water delivered, to the total water diverted or pumped into an open channel or pipeline at the upstream end.

irrigation efficiency (E_i, IE)
ratio of the average depth of irrigation water that is beneficially used to the average depth of irrigation water applied.

motor efficiency(E_m)
ratio of the power delivered to the pump by the power unit to the input power to the motor.

project efficiency
overall efficiency of irrigation water use in a project setting that accounts for all water uses and losses, such as crop ET, environmental control, salinity control, deep percolation, runoff, ditch and canal leakage, phreatophyte use, wetlands use, operational spills, and open water evaporation.

pump efficiency (E_p)
ratio of the water power produced by the pump, to the power delivered to the pump by the power unit

(water) storage efficiency
ratio of the amount of water stored in the root zone during irrigation to the amount of water needed to fill the root zone to field capacity.

water use efficiency (WUE)
ratio of the yield per unit area to the applied irrigation water per unit area.

- **efficiency standard**

a value or criteria that establishes levels or conditions of water use in the ornamental landscape.

- **effluent**

any substance, particularly a liquid, that enters the environment from a point source. Generally refers to wastewater from a sewage treatment or industrial plant.

- **effluent irrigation**

land application of wastewater for irrigation and beneficial use of nutrients.

- **El Niño/La Niña**
a transient, periodic warming of the equatorial eastern Pacific Ocean, associated with fisheries depletion and large-scale climatic fluctuations.

- **electrical conductivity**
measure of the ability of the soil water to transfer an electrical charge. Used as an indicator for the estimation of salt concentration. (EC_{e}) Electrical conductivity of soil water extract. (EC_{i}) Electrical conductivity of irrigation water. (ECa_{w}) Electrical conductivity of applied water.

- **electrical resistance block**
block made up of various materials containing electrical contact wires that is placed in the soil at selected depths to measure soil moisture content (tension). Electrical resistance, as affected by moisture in the block is read with a meter.

- **electrodialysis**
a process which uses an electrical current and an arrangement of permeable membranes to separate soluble minerals from water. It is often used to desalinate salt or brackish water.

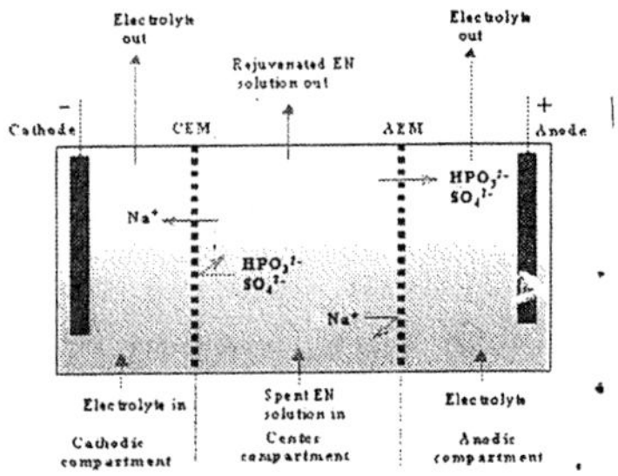

- **elevation head**
a measurement of pressure.

- **ell**
a fitting used to change the direction of a pipe. For PVC and poly irrigation uses they are available in 90 degree and 45 degree bends. Some specialty ells are available in other angles but have limited availability. You can remember ells simply by their shape, they look like an 'L'. PVC ells are available with threads in both ends, threads in one end and a glued socket in the other, or with glue sockets in both ends. Insert ells come with male threads and barbs, female threads and barbs, glue sockets and barbs, glue spigots and barbs or barbs and barbs.

- **emission point**
  location where water is discharged from an emitter.
- **emission uniformity (EU)(-,%)**
  index of the uniformity of emitter discharge rates throughout a micro-irrigation system. Takes account of both variations in emitters and variations in the pressure under which the emitters operate.
- **emitter**
  small micro-irrigation dispensing device designed to dissipate pressure and discharge a small uniform flow or trickle of water at a constant discharge, which does not vary significantly because of minor differences in pressure head. Also called a 'dripper' or 'trickler'. compensating emitter.
  continuous flushing emitter - micro irrigation system emitter designed to continuously permit passage of large solid particles while operating at a trickle or drip flow, thus reducing filtration.
  requirements flushing emitter- emitter designed to have flushing flow of water to clear the discharge opening every time the system is turned on.
  line source emitter - water is discharged from closely spaced perforations, emitters or a porous wall along the tubing.
  long-path emitter- emitter which employs a long capillary sized tube or channel to dissipate pressure.
  multi-outlet emitter- device which supplies water to two or more points through small diameter auxiliary tubing.
  non-pressure compensating emitter- emitter designed with a fixed orifice or other components and contains no pressure compensating features. ***orifice emitter***- emitter which employs a series of orifices to dissipate pressure.
  pressure compensating emitter – emitter designed to discharge water at a near constant rate over a wide range of lateral line pressures.
  vortex emitter – emitter which employs a vortex effect to dissipate pressure.

- **endangered species**
  a species in imminent danger of extinction.

- **energy**
  the product of power and time, measured in Watt-hours. 1000 Watt-hours = 1 KiloWatt-hour (abbreviation: kWh). Variation: the product of current and time is Ampere-hours, also called Amp-hours (abbreviation: Ah). 1000 Watt consumed for 1 hour = 1 kWh.

- **energy budget**
  an accounting of all energy inputs and outputs for a system.

- **energy conservation**
  using energy resources in such a way as to minimise energy consumption in relation to benefits gained.

- **energy efficiency**
  the amount of utility, either work performed or income generated, gained per unit of an energy resource.

- **enteric viruses**
  a category of viruses related to human excreta found in waterways.

- **entrance loss**
  energy lost in eddies and friction at the inlet to a conduit or structure.

- **environment**
  aggregate of external conditions that influence the life of an individual organism or population.

- **environmental cognition**
  the mental process of making sense of the world that each of us inhabits.

- **environmental ethics**
  a philosophical position regarding the relation between humans and nature.

- **ephemeral**
  a stream which flows only in direct response to precipitation, and whose channel is at all times above the water table.

- **epilimnion**
  warm, less dense top layer in a stratified lake. Compare hypolimnion. See **thermal stratification**

- **equity**
  fairness in the use and allocation of resources.

- **erosion**
  the wearing away of the land surface by wind, water, ice or other geologic agents. Erosion occurs naturally from weather or runoff but is often intensified by human land use practices.

- **erosivity**
  the ability of rainfall to cause erosion. Erosivity is a function of rainfall intensity and drop size.

- **estuarine waters**
  deepwater tidal habitats and tidal wetlands that are usually enclosed by land but have access to the ocean and are at least occasionally diluted by freshwater runoff from the land (such as bays, mouths of rivers, salt marshes, lagoons).

- **estuarine zone**
  area near the coastline that consists of estuaries and coastal saltwater wetlands.

- **estuary**
  a semi-enclosed water body, open to the sea, in which seawater is significantly diluted by fresh water from the land.

- **et**
  the amount of water lost by plant foliage through transpiration and the surrounding soil through evaporation which must be replaced by irrigation and precipitation. See **water budget**.

- **eutrophication (natural)**
  an excess of plant nutrients from natural erosion and runoff from the land in an aquatic ecosystem supporting a large amount of aquatic life that can deplete the oxygen supply.

- **evaporation**
  the process by which water is changed from the liquid or the solid state into the vapour state. In hydrology, evaporation is vaporisation that takes place at a temperature below the boiling point.

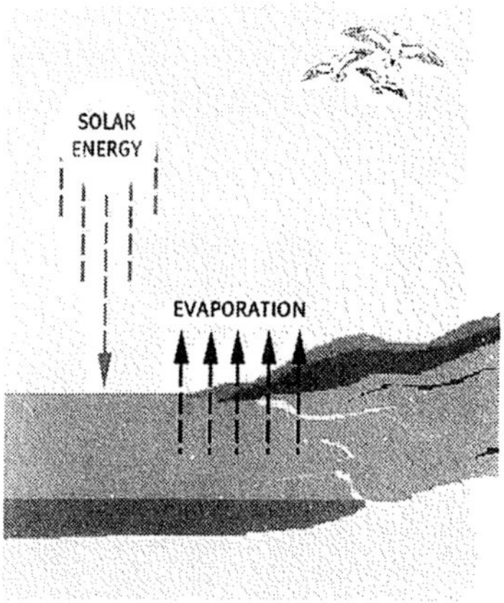

- **evaporation opportunity (relative evaporation)**
  the ratio of the rate of evaporation from a land or water surface in contact with the atmosphere, to the evaporativity under existing atmospheric conditions. It is the ratio of actual to potential rate of evaporation, generally stated as a percentage. The opportunity for a given rate of evaporation to continue is determined by the available moisture supply.

- **evaporation pan**
  an open tank used to contain water for measuring the amount of evaporation.

- **evaporation pond**
  a containment area where liquids are allowed to evaporate. In some cases a spraying mechanism is used to speed evaporation.

- **evaporation, total**
  the sum of water lost from a given land area during any specific time by transpiration from vegetation and building of plant tissue; by evaporation from water surfaces, moist soil, and snow; and by interception. It has been variously termed 'evaporation, 'evaporation from land areas,' 'evapotranspiration,' 'total loss,' 'water losses.'

- **evaporativity (potential rate of evaporation)**
  the rate of evaporation under the existing atmospheric conditions from a surface of water that is chemically pure and has the temperature of the atmosphere.

- **evapotranspiration**
  combination of water transpired from vegetation and evaporated from the soil and plant surfaces.
  crop (ET_c)
  crop evapotranspiration is the quantitative amount of ET within the cropped area of a field, and which is associated with growing of a

crop. Same as plant water requirement.
Amount of water used by the crop in transpiration and building of plant tissue, and that evaporated from adjacent soil or intercepted by plant foliage. It is sometimes referred to as consumptive use.

potential (ET_p, E_{tp})
Rate at which water, if available, would be removed from soil and plant surfaces rate at which water, if available, would be removed from the soil and plant surface expressed as the latent heat transfer per unit area or its equivalent depth of water per unit area.

reference
[ET_o]
rate of evapotranspiration from an extensive surface cool-season green grass cover of uniform height of 12 cm, actively growing, completely shading the ground, and not short of water.
[ET_r]
upper limit or maximum evapotranspiration that occurs under given climate conditions with a field having a well-watered agricultural crop with an aerodynamically rough surface,

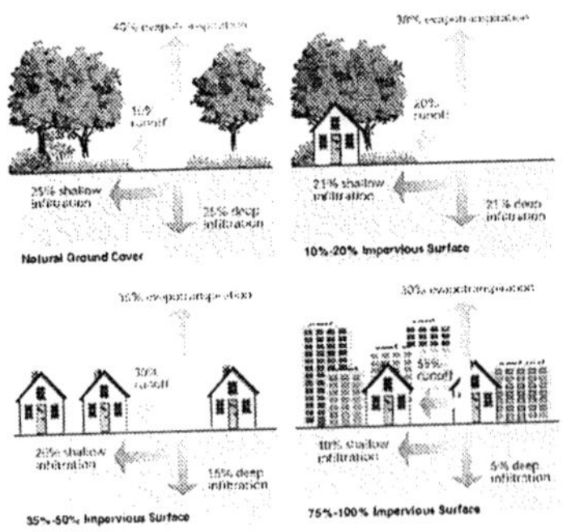

- **evergreen**
  a woody plant that never loses all its foliage at one time. Foliage is retained year-round.

- **excessive rainfall**
  see **rainfall, excessive**
- **exchange capacity**
  see **field-moisture capacity**

- **exploitation**
use of a resource at the maximum profitable short-term rate, without regard for long-term resource quality or availability.

- **exposure**
a combination of the aspect - north, south, east, west - and the protection - walls, fences, other plants - influencing a plant's habitat. Exposure significantly influences the survival and growth behaviour of a plant. See **micro-climate**.

- **externality**
a non-market exchange, in which at least one party to the exchange is not compensated and may have little choice in the exchange.

- **extinction**
complete disappearance of a species because of failure to adapt to environmental change.

- **fallow**
land ploughed and tilled and left unplanted.

- **farm loss (water)**
water delivered to a farm which is not made available to the crop to be irrigated.

- **farmland**
land that is part of farm units, including cropland, pasture, small woodlots and areas used for small farm roads and buildings.

- **faecal coliform**
the portion of the coliform bacteria group which is present in the intestinal tracts and faeces of warm-blooded animals. A common pollutant in water.

- **female adapter**
a fitting used to adapt from solvent welded PVC to a threaded or barbed connection.

- **fertigation**
application of nutrients through an irrigation system.

- **fertilise**
in flowers, it refers to pollination; to fertilise a plant is to apply fertiliser.

- **fertiliser**
a substance added to the soil to improve plant growth. The most commonly used fertilisers are those containing large amounts of nitrogen, potassium and phosphorus.

• **FHT**
female hose thread (as found on the beginning of a garden hose).

• **field capacity**
is the amount of water in the soil after all free gravitational water has been removed (drained via gravity) and no additional water can be held in the soil. See **field-moisture capacity**.

• **field-moisture capacity**
the quantity of water which can be permanently retained in the soil in opposition to the downward pull of gravity.

• **field-moisture deficiency**
the quantity of water, which would be required to restore the soil moisture to field-moisture capacity.

• **fill slope**
describes the process in landscaping of levelling a slope by filling and adding retaining walls to level an area. Attention to compacting the soil in filled area is critical. See **cut slope**.

• **filter**
a device used to remove solids from a mixture or to separate materials. Materials are frequently separated from water using filters.

• **filtration**
the mechanical process which removes particulate matter by separating water from solid material, usually by passing it through sand.

• **final infiltration rate**
see **basic intake rate**

• **fine sand**
soil textural class

• **fine sandy loam**
soil textural class. See also **soil texture**.

• **FIPT**
acronym for female iron pipe thread.

• **fittings**
the generic name for the various parts that attach the pipes together. Includes bushings, couplings, cro-

sses, ells, female adapters, male adapters, reducers and tees. Fittings may be threaded, barbed, soldered or welded to the pipe.

- **fixed amount-frequency scheduling**
  method of irrigation scheduling that involves water delivery at a fixed rate or a fixed volume and at constant intervals. Examples include rotation and continuous flow methods. Considered a rigid form of scheduling.

- **fixed costs**
  costs of operating a business that do not vary with the rate of output of goods and services.

- **fixed ground water**
  water held in saturated material with pore spaces so small that it is permanently attached to the pre-walls or moves so slowly that it is usually not available as a source of water for pumping.

- **flatwoods (pine)**
  natural communities that occur on level land and are characterised by a dominant overstory of slash pine. Depending upon soil drainage characteristics and position in the landscape, pine flatwoods habitats can exhibit xeric to moderately wet conditions.

- **flocculation**
  large scale treatment process involving gentle stirring whereby small particles in flocs are collected into larger particles so their weight causes them to settle to the bottom of the treatment tank.

- **flood**
  an overflow or inundation that comes from a river or other body of water, and causes or threatens damage.
  Any relatively high streamflow overtopping the natural or artificial banks in any reach of a stream.
  A relatively high flow as measured by either guage height or discharge quantity. *See* **annual flood**

- **flood control pool**
  reservoir volume reserved for flood runoff and then evacuated as soon as possible to keep that volume in readiness for the next flood.

- **flood crest**
  see **flood peak**

- **flood event**
  see **flood wave**

- **flood irrigation**
  method of irrigating where water is applied from field ditches onto land which has no guide preparation such as furrows, borders or corrugations.

- **flood peak**
  the highest value of the stage or discharge attained by a flood, thus, peak stage or peak discharge. Flood crest has nearly the same meaning, but since it connotes the top of the flood wave, it is properly used only in referring to stage—thus, crest stage, but not crest discharge.

- **flood plain**
  a strip of relatively smooth land bordering a stream, built of sediment carried by the stream and dropped in the slack water beyond the influence of the swiftest current. It is called a living flood plain, if it is overflowed in times of highwater; but a fossil flood plain if it is beyond the reach of the highest flood. The lowland that borders a river, usually dry but subject to flooding. That land outside of a stream channel described by the perimeter of the maximum probable flood.

- **flood profile**
  a graph of elevation of the water surface of a river in flood, plotted as ordinate, against distance, measured in the downstream direction, plotted as abscissa. A flood profile may be drawn to show elevation at a given time, crests during a particular flood or to show stages of concordant flows.

- **flood routing**
  the process of determining progressively the timing and shape of a flood wave at successive points along a river.

- **flood stage**
  the gauge height of the lowest bank of the reach in which the gauge is situated. The term 'lowest

bank' is, however, not to be taken to mean an unusually low place or break in the natural bank through which the water inundates an unimportant and small area. It is the stage at which overflow of the natural banks of a stream begins to cause damage in the reach in which the elevation is measured.

- **flood wave**
  a distinct rise in stage culminating in a crest and followed by recession to lower stages.

- **flood zone**
  the land bordering a stream which is subject to floods of about equal frequency; for example, a strip of the flood plain subject to flooding more often that once but not as frequently as twice in a century.

- **flood, 100-year**
  a 100-year flood does not refer to a flood that occurs once every 100 years, but to a flood level with a 1 percent chance of being equalled or exceeded in any given year.

- **flood, maximum probable**
  the largest flood for which there is any reasonable expectancy in this climatic era.

- **flood-control storage**
  storage of water in reservoirs to abate flood damage.

- **flood-frequency curve**
  1. a graph showing the number of times per year on the average, plotted as abscissa, that floods of magnitude, indicated by the ordinate, are equalled or exceeded.
  2. a similar graph but with recurrence intervals of floods plotted as abscissa.

- **floods above a base**
  see **partial-duration flood series**

- **floodway**
  a part of the flood plain. The channel of a river or stream; the parts of the flood plains adjoining the channel which are reasonably required to carry and discharge the flood water or floodflow of any river or stream.

- **flora**
  plant population of a region.

• **flow**
the rate of water discharged from a source expressed in volume with respect to time.

• **flow augmentation**
the addition of water to meet flow needs.

• **flow rate**
(Q or q)(gpm, gph, gal/min, ft^3/s, cfs)
Rate of flow or volume per unit period of time.

• **flow resource**
a resource that is simultaneously used and replaced. Perpetual and renewable resources are flow resources.

• **flow-duration curve**
a cumulative frequency curve that shows the percentage of time that specified discharges are equalled or exceeded.

• **flowing well/spring**
a well or spring that taps ground water under pressure so that water rises without pumping. If the water rises above the surface, it is known as a flowing well.

• **flowing wells**
a well where the piezometric surface lies above the ground surface.

• **flushing emitter**
designed to have a flushing flow of water to clear the discharge opening every time the system is turned on.

• **food security**
the condition of having both physical and economic access to the basic food that people need to function normally.

• **foot valve**
a check valve placed in the water source below a surface pump. It prevents water from flowing back down the pipe and 'losing prime'.

• **forbay**
the water behind a dam.

• **forest influences**
effects resulting from the presence of forest or brush upon climate, soil water, runoff, streamflow, floods, erosion and soil productivity.

- **FPT**
  female pipe thread.

- **frazil (frazil ice).**
  a French-Canadian term for fine spicular ice, derived from the French for cinders which this variety of ice most resembles. When formed in salt water, it is known as lolly ice. It is composed of fine particles which, when first formed, are colloidal and not seen in the water in which they are floating.

- **free drainage**
  movement of water by gravitational forces through and below the plant root zone. This water is unavailable for plant use except while passing through the soil.

- **free ground water**
  water in interconnected pore spaces in the zone of saturation down to the first impervious barrier, moving under the control of the water table slope.

- **freezing**
  the change of a liquid into a solid as temperature decreases. For water, the freezing point is 32°F or 0°C.

- **frequency demand scheduling**
  method of irrigation scheduling similar to demand scheduling, but typically involves a fixed duration of the delivery, such as 24 hours. This method is considered flexible, although somewhat less so than demand scheduling from the water users' perspective.

- **fresh salt water interface**
  the region where fresh water and salt water meet. In the Edwards region, it is commonly referred to as the 'bad water line', although it is zone and not a line.

- **freshwater**
  water that contains less than 1,000 milligrams per litre (mg/L) of dissolved solids, generally, more than 500 mg/L of dissolved solids is undesirable for drinking and many industrial uses.

- **friable**
  soil consistency term referring to the ease with which the soil aggregates may be crumbled (in the hand), i.e. a friable soil is easily crumbled in the hand.

- **friction factor (lateral)**
  factor used to size pipe.

- **friction factor, Christiansen (F)(-)**
  factor used to account for multiple outlets on a lateral pipe when computing friction loss via the Christiansen Procedure.

- **friction factor, Darcy Weisbach** (f)(-)
  friction factor used with Darcy Weisbach equation.

- **friction loss(h_f(psi, ft)**
  also, referred to a pressure loss.
  Amount of pressure lost through pipes due to water movement and turbulence. Amount of pressure lost as water flows through an irrigation system (due to friction against the pipe walls). As water moves through an irrigation system, pressure is lost because of turbulence created by the moving water. This turbulence can be created in pipes, valves or fittings. These losses are referred to as friction losses.

- **frost**
  a covering of minute ice crystals on a cold surface.

- **frost protection**
  applying irrigation water to affect air temperature, humidity, and dew point to protect plant tissue from freezing. The primary source of heat (called heat of fusion) occurs when water turns to ice, thus protecting sensitive plant tissue.

- **full irrigation**
  management of water applications to fully replace water used by plants over an entire field.

- **fungicide**
  chemical pesticide that kills fungi, or prevents them from causing diseases on plants.

- **furrow**
  small channel for conveying irrigation water down slope across the field. Sometimes referred to as a rill or corrugation.

- **furrow irrigation**
  a type of irrigation in which water is allowed to flow along the furrows (troughs) between rows of crops.

- **furrow stream**
  stream flow in a furrow, corrugation or rill.

- **gauge**
device for registering water level, discharge, velocity, pressure, etc.

- **gauge height**
the water-surface elevation referred to some arbitrary gauge datum. Gauge height is often used interchangeably with the more general term stage although gauge height is more appropriate when used with a reading on a gauge.

- **gauging station**
a particular site on a stream, canal, lake or reservoir where systematic observations of gauge height or discharge are obtained.

- **gaining**
a stream or reach of a stream that receives water from the zone of saturation. An effluent stream.

- **gallon**
a unit of volume.

- **gallons per minute**
a measurement of water flow.

- **gate (irrigation)**
structure or device for controlling the rate of flow into or from a canal or ditch.

- **gate valve**
refers to the operating mechanism for the valve, which is a sliding gate which moves up or down to block the flow. Often used as isolation valves. Never used as control valves. Because the gate slides it is very subject to wear, and gate valves wear out fast when used often. Some gate valves use a wedge-shaped gate which holds up better. They are still not designed for regular use, but for emergency shut-off only. Use a cheap gate valve and you will soon learn to 'hate the gate'.

- **gate, slide gate**
device used to control the flow of water to, from or in a pipeline or open channel.

- **gated pipe**
portable pipe with small gates installed along one side for distributing irrigation water to corrugations or furrows.

- **gear drive sprinkler**
see **sprinkler head.**

- **general systems theory**
  a way of looking at the world or any part of it as an interacting set of parts.
- **generational equity**
  the fairness doctrine applied to subsequent generations so that they receive the environment in the same or better condition than the generation before them.
- **geographic information system (GIS)**
  a computer database and data-manipulation system designed to use geographically organised data. A map based database plots locations of information on maps using latitude and longitude.
- **geohydrology**
  a term which denotes the branch of hydrology relating to subsurface or subterranean waters, that is, to all waters below the surface.
- **geologic erosion**
  normal or natural erosion caused by geological processes acting over long geologic periods and resulting in the wearing away of mountains, the building up of floodplains, coastal plains, etc.
- **geologic estimate of resource**
  an estimate of the amount of a mineral resource in the earth based on information about the concentration and distribution of that mineral in rocks, without regard for the economics of extraction.
- **geopressured reservoir**
  a geothermal reservoir consisting of porous sands containing water or brine at high temperature or pressure.
- **geopressurised resource**
  a geothermal resource in which hot groundwater is pressurised by natural forces.
- **geothermal energy**
  energy extracted from heat

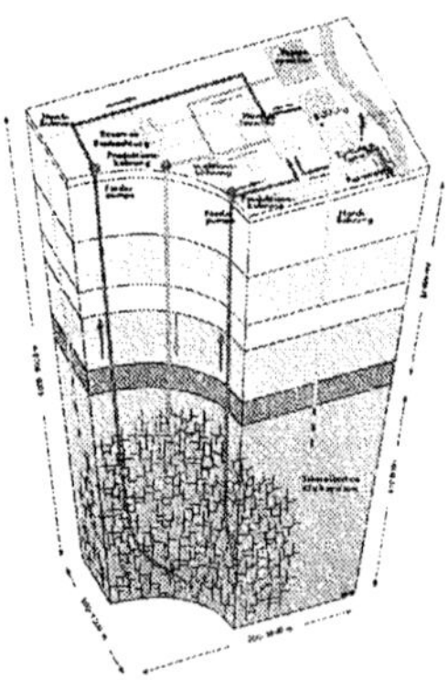

contained in rocks near the earth's surface.

- **geyser**
a geothermal feature of the Earth where there is an opening in the surface that contains superheated water that periodically erupts in a shower of water and steam.

- **giardiasis**
a disease that results from an infection by the protozoan parasite Giardia Intestinalis, caused by drinking water that is either not filtered or not chlorinated. The disorder is more prevalent in children than in adults and is characterised by abdominal discomfort, nausea, and alternating constipation and diarrhoea.

- **glacier**
bodies of land ice that consist of recrystallised snow accumulated on the surface of the ground

- **globe valve**
refers to the water flow pattern into and out of the valve. Often used as control valves. Seldom used as isolation valves. The valve inlet is on one side of the globe valve and the outlet is on the other side. Globe valves as a group tend to be very reliable, but have slightly higher friction losses than 'angle' valves, the other common style used for control valves.

- **GPD**
gallons per day

- **GPH**
refers to water flow; gallons per hour.

- **GPM**
gallons per minute

- **grab sample**
a sample taken at a given place and time. Compare **composite sample**.

- **grading**
the process of adjusting the slope in a landscape area. Grading is 'rough' to attain the general desired slope for drainage and 'fine' to smooth out the surface for hardscape and plant materials.

- **granular activated carbon**
  pure carbon heated to promote 'active' sites which can adsorb pollutants. Used in some home water treatment systems to remove certain organic chemicals and radon.

- **grassland**
  a biome dominated by grasses. Most grasslands have semiarid climates.

- **gravitational water**
  soil water that moves into, through, or out of the soil under the influence of gravity.

- **gravity (acceleration due to)**
  acceleration caused by the attraction of the mass of earth to bodies at or near its surface. (i.e. 32.2 ft/sec^2)

- **gravity flow**
  the term given a water system that relies on gravity to provide the pressure required to deliver the water. Consists of a water source located at a higher elevation than the water delivery points.

- **gravity irrigation**
  irrigation method that applies irrigation water to fields by letting it flow from a higher level supply canal through ditches or furrows to fields at a lower level.

- **green revolution**
  a variety of agricultural systems developed for application in developing countries, involving the introduction of improved seed varieties, fertilisers, and irrigation systems.

- **greenhouse effect**
  the tendency of the atmosphere to be transparent to shortwave solar radiation but opaque to longwave terrestrial radiation, leading to a warming of the atmosphere.

- **greenhouse gases**
  substances that are transparent to shortwave (solar) radiation but absorb longwave (terrestrial) radiation and thus contribute to warming of the atmosphere. Carbon dioxide, ozone, chlorofluorocarbons, methane, and water

vapour are important greenhouse gases.

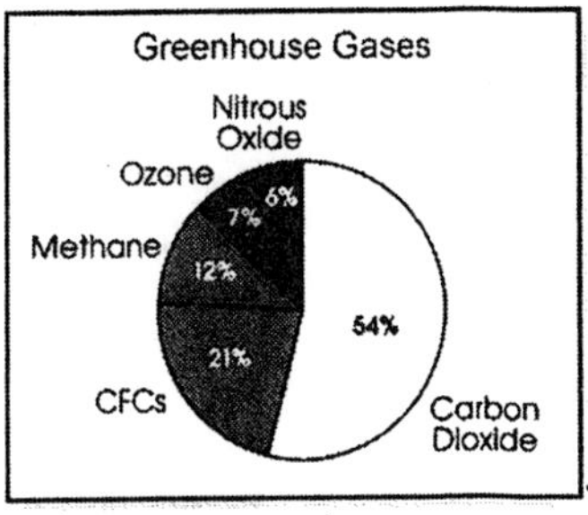

- **grey water**
  wastewater from clothes washing machines, showers, bathtubs, hand-washing, lavatories and sinks that are not used for disposal of chemical or chemical-biological ingredients.

- **gross irrigation**
  see **irrigation, gross.**

- **gross irrigation system capacity**
  see **system capacity, gross irrigation**

- **gross water (irrigation) requirement**
  see **irrigation requirement, gross.**

- **ground bark**
  the bark of trees shredded or ground up for use as mulch.

- **ground fire**
  a forest fire that only burns at ground level, consuming litter and downed trees but not live standing trees.

- **ground water**
  1. water that flows or seeps downward and saturates soil or rock, supplying springs and wells. The upper surface of the saturated zone is called the water table.
  2. water stored underground in rock crevices and in the pores of geologic

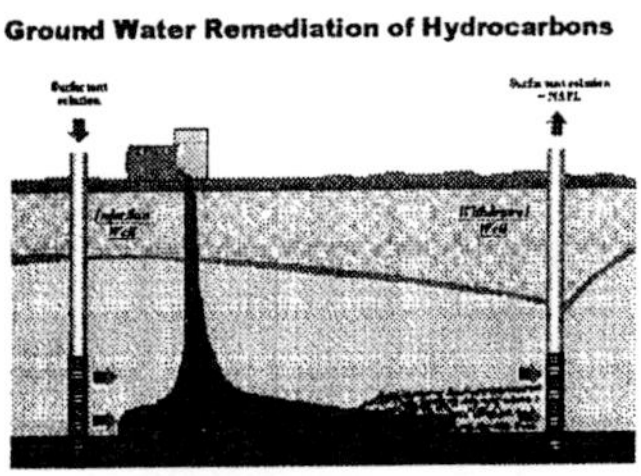

materials that make up the Earth's crust.

- **ground water hydrology**
  the branch of hydrology that deals with ground water; its occurrence and movements, its replenishment and depletion, the properties of rocks that control ground water movement and storage and the meth-

ods of investigation and utilisation of ground water.

- **ground water recharge**
  the inflow to a ground water reservoir.

- **ground water registration**
  a statement made by a well owner registering the beneficial use of ground water.

- **ground water reservoir**
  an aquifer or aquifer system in which ground water is stored. The water may be placed in the aquifer by artificial or natural means.

- **ground water runoff**
  a portion of runoff which has passed into the ground, has become ground water and has been discharged into a stream channel as spring or seepage water.

- **ground water storage**
  the storage of water in ground water reservoirs.

- **ground water, confined**
  ground water under pressure significantly greater than atmospheric, with its upper limit the bottom of a bed with hydraulic conductivity distinctly lower than that of the material in which the confined water occurs.

- **ground water, unconfined**
  water in an aquifer that has a water table that is exposed to the atmosphere.

- **groundwater mining (overdraft)**
  pumping of groundwater for irrigation or other uses, at rates faster than the rate at which the groundwater is being recharged.

- **ground-water outflow**
  that part of the discharge from a drainage basin that occurs through the ground water. The term 'underflow' is often used to describe the ground-water outflow that takes place in valley alluvium (instead of the surface channel) and thus is not measured at a gauging station.

- **ground-water recharge**
  inflow of water to a ground-water reservoir from the surface. Infiltration of precipitation and its movement to the water table is one form of natural recharge. Also, the volume of water added by this process.

- **groundwater reservoir**
  an aquifer or aquifer system in which ground water is stored. The water may be placed in the aquifer by artificial or natural means.

- **groundwater table**
  the upper boundary of groundwater where water pressure is equal to atmospheric pressure, i.e., water level in a bore hole after equilibrium when groundwater can freely enter the hole from the sides and bottom.

- **growing season**
  the period, often the frost-free period, during which the climate is such that crops can be produced.

- **guest worker**
  a person allowed in a country on a temporary basis in order to increase the available labour force in that country.

- **gully**
  a steep-walled stream channel incised in the soil by accelerated erosion. A deeply eroded channel caused by the concentrated flow of water.

- **gully reclamation**
  use of small dams of manure and straw, earth, stone, or concrete to collect silt and gradually fill in channels of eroded soil.

- **guttation**
  the loss of water in liquid form from the uninjured leaf or stem of the plant, principally through water stomata.

- **gypsum block**
  electrical resistance block in which the material used to absorb water is gypsum. It is used to measure soil water content in non-saline soils.

- **habitat**
  land that provides living space and sustenance for plants and animals.

- **hail**
  a form of precipitation which forms into balls or lumps of ice over 0.2 inch in diameter. Hail is formed by alternate freezing and melting as precipitation is carried up and down in highly turbulent air currents.

- **halocline**
  a marked change in salinity at a particular depth in the ocean or an estuary; it sig-

nals the boundary between two layers of water.

- **hard water**
  water containing a high level of calcium, magnesium and other minerals. Hard water reduces the cleansing power of soap and produces scale in hot water lines and appliances.

- **hardness**
  a water-quality indication of the concentration of alkaline salts in water, mainly calcium and magnesium. If the water you use is 'hard' than more soap, detergent or shampoo is necessary to raise a lather.

- **hardpan**
  a shallow layer of earth material which has become relatively hard and impermeable, usually through the deposition of minerals.

- **hardwoods**
  trees with particularly dense wood, primarily broad-leafed trees.

- **hardy**
  capability of a plant to survive in a given environment.

- **harvested rainwater**
  the rain that falls on a roof or yard and is channelled by gutters or channels to a storage tank. The first wash of water on a roof is usually discarded and the subsequent rainfall is captured for use if the system is being used for potable water. Good quality water is available by this method in most areas.

- **head**
  the pressure of a fluid owing to its elevation, usually expressed in feet of head or in pounds per square inch, since a measure of fluid pressure is the height of a fluid column above a given or known point.

- **head ditch**
  ditch across the upper end of a field used for distributing water in surface irrigation.

- **head gate**
  water control structure at the entrance to a conduit or canal.

- **head spacing**
  see **spacing between sprinklers**.

- **head to head**
  the movement pattern of a seasick cruise ship passenger. In irrigation 'head to head' refers to the situation where sprinklers are spaced so that the water from one sprinkler throws all the way to the next sprinkler. Most sprinklers are designed to give the best performance when head to head spacing is used.

- **headgate**
  the gate that controls water flow into irrigation canals and ditches. A watermaster regulates the headgates during water distribution and posts headgate notices declaring official regulations.

- **headwater(s)**
  1. the source and upper reaches of a stream; also the upper reaches of a reservoir.
  2. the water upstream from a structure or point on a stream.
  3. the small streams that come together to form a river. Also may be thought of as any and all parts of a river basin except the mainstream river and main tributaries.

- **heat budget, annual (of a lake).**
  the amount of heat necessary to raise the water from the minimum temperature of winter to the maximum temperature of summer

- **heat of vaporisation**
  the amount of heat necessary to convert a liquid (water) into vapour.

- **heavy soil**
  see **clay soil.**

- **heavy water**
  water in which all the hydrogen atoms have been replaced by deuterium.

- **herbaceous**
  a plant whose stems and leaves die to the ground each year. All non-woody plants including bulbs, rhizomes, corms, tubers,

annuals and herbaceous perennials. See **herbaceous perennial**.

- **herbaceous perennial**
  a plant whose stems and leaves die to the ground each year and it regrows stems and foliage the following growing season. See **perennial.**

- **herbicide**
  chemical substance designed to kill or inhibit the growth of plants, especially weeds.

- **High density polyethylene (HDPE)**
  one of several forms of polyethylene used to make pipe and other irrigation components.

- **highest and best use**
  the classification of water based on an analysis of the greatest needs of the future. Certain quantities are reserved for appropriation according to the classification.

- **holding pond**
  a small basin or pond designed to hold sediment laden or contaminated water until it can be treated to meet water quality standards or be used in some other way.

- **homocentric**
  a view of nature that only considers human, rather than plant or animal, needs.

- **hopper-burn**
  the bronzing of leaves caused by leafhoppers injecting toxins into the leaf while feeding.

- **horizon (soil)**
  layer of soil or soil material approximately parallel to the land surface and differing from adjacent genetically related layers in physical, chemical and biological properties or characteristics such as colour, structure, texture, consistency, kinds and number of organisms present, degree of acidity or alkalinity, etc.

- **horsepower water (Whp, WHP)**
  Energy added to water by a pump.
  input horsepower (Ihp, IHP)
  Energy added to a motor or engine
  brake horsepower *(Bhp, BHP)
  power required to drive a pump.

- **hose bib**
valve configured to be mounted on a wall having threads to accommodate the connection of a water hose.

- **hose drag traveller**
see **irrigation system**

- **humid climate**
see **climate.**

- **humus**
decomposed organic matter of soil that no longer resembles the vegetable and animal matter from which it originated. Often incorrectly referred to as any decomposing organic material including sawdust, ground bark and manure.

- **hydrant**
outlet, usually portable, used for connecting surface irrigation pipe to an alfalfa valve outlet.

- **hydraulic conductivity**
coefficient describing the ease at which the soil pores permit water movement. Soil-water characteristic describing the ability of water to flow through a particular soil.

- **hydraulic efficiency**
efficiency of a pump or turbine to impart energy to or extract energy from water. The ability of hydraulic structure or element to conduct water with minimum energy loss.

- **hydraulics**
the science of fluids in motion. As it relates to irrigation there are two sub-categories for hydraulics. They are open-channel hydraulics (canals, ditches, streams, rivers, etc.) and closed channel (or closed conduit) hydraulics (pipelines, tanks, etc.).

- **hydro**
the prefix denoting water or hydrogen.

- **hydroelectric**
having to do with production of electricity by water power from falling water.

- **hydroelectric plant**
electric power plant in which the energy of falling water is used to spin a turbine generator to produce electricity.

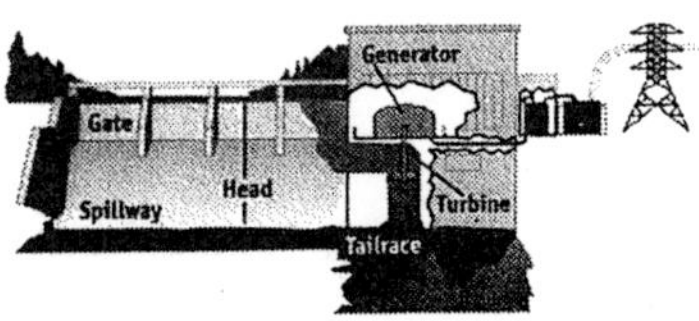

- **hydroelectric power water use**
  the use of water in the generation of electricity at plants where the turbine generators are driven by falling water.

- **hydrograph**
  a graph showing stage, flow, velocity or other property of water with respect to time.

- **hydrologic budget**
  an accounting of the inflow to, outflow from, and storage in, a hydrologic unit, such as a drainage basin, aquifer, soil zone, lake, reservoir or irrigation project.

- **hydrologic cycle**
  a convenient term to denote the circulation of water from the sea, through the atmosphere, to the land; and thence, with many delays, back to the sea by overland and subterranean routes, and in part by way of the atmosphere; also the many short circuits of the water that is returned to the atmosphere without reaching the sea.

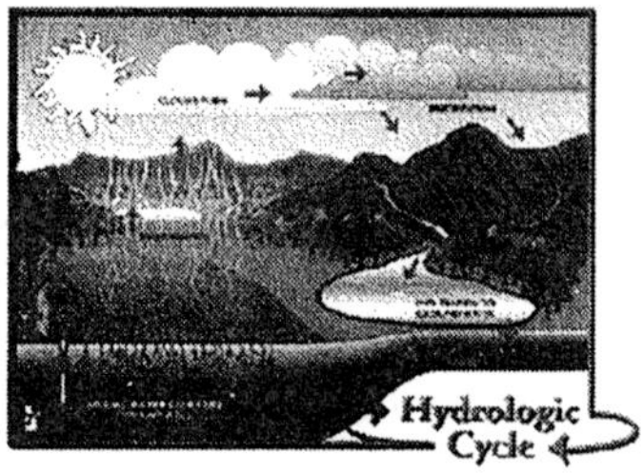

- **hydrologic equation**
  the equation balancing the hydrologic budget.

- **hydrologic unit**
  is a geographic area representing part or all of a surface drainage basin or distinct hydrologic feature.

- **hydrology**
  the science encompassing the behaviour of water as it occurs in the atmosphere, on the surface of the ground, and underground. The science that relates to the water of the earth. The science treating of the waters of the earth, their occurrence, distribution, and movements. In practice the study of the water of the oceans and the atmosphere is considered part of the sciences of oceanography and meteorology.

- **hydrometer**
  an instrument used to measure the density of a liquid.

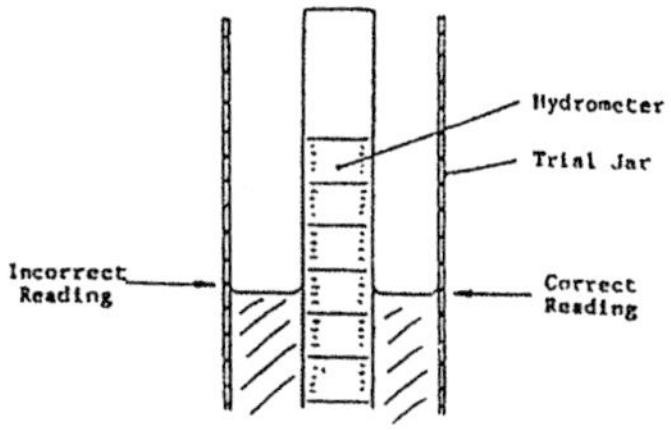

- **hydroperiod**
  the frequency and duration of inundation or saturation of an ecosystem. In the context of characterising wetlands, the term hydroperiod describes that length of time during the year that the substrate is either saturated or covered with water.

- **hydropower**
  electrical energy produced by falling water.

- **hydrosphere**
  region that includes all the earth's liquid water, frozen water, floating ice, frozen upper layer of soil, and the small amounts of water vapour in the Earth's atmosphere.

- **hydrostatic head**
  a measure of pressure at a given point in a liquid in terms of the vertical height of a column of the same liquid which would produce the same pressure.

- **hydrostatic pressure**
  pressure exerted by or existing within a liquid at rest with respect to adjacent bodies.

- **hydrothermal mineralisation**
  a process of concentration of metallic ores caused by high-temperature geochemical processes in underground waters.

- **hydrozone**
  1. a landscape area or portion of a landscape area having plants with similar water requirements. The area may or may not require irrigation. Also referred to as water zone.
  2. an area of an irrigation system where all the factors that influence the watering schedule are similar. Typical factors to be considered would be the type of plants, the precipitation rate of sprinklers or emitters, solar radiation, wind, soil type, and slope. See the related term **valve zone**.

- **hyetograph**
  graphical representation of rainfall intensity against time.

- **hygroscopic nuclei**
piece of dust or other particle around which water condenses in the atmosphere. These tiny droplets then collide and coalesce, with as many as 10,000 nuclei contributing to formation of a raindrop.

- **hygroscopic water**
1. water which is bound tightly by the soil solids at potential values lower than –31 bars.
2. water that is tightly held by soil particles. It does not move with the influence of capillary action or gravity, and it is normally unavailable to plants.

- **hypolimnion**
bottom layer of cold water in a lake. See **thermal stratification**

- **hysteresis**
is an effect on the soil water content that occurs between wetting and drying. If a soil is becoming wet due to irrigation or precipitation, it will increase in water content as the soil water tension (how strongly the water is held in the soil) decreases. As drying (water removal) occurs, the tension will increase as the water is removed. For soils with strong hysteresis characteristics (loams and clays in particular), the soil water content will not be the same during drying as it was during wetting, even though the soil water tension is the same. This effect is related to the pore space in the soil and how the soil water fills those pores while wetting and then releases water while drying.

- **ice**
a solid form of water.

- **ID**
abbreviation for inside diameter (usually of a pipe).

- **impact power (watts, horsepower)**
rate at which drops deliver kinetic energy to the soil. Kinetic energy of impact on soil per unit volume.

- **impact rate**
impact power per unit area. Impact power per square foot.

- **impermeable**
material that does not permit fluids to pass through.

- **impermeable layer**
  a layer of solid material, such as rock or clay, which does not allow water to pass through.
- **impervious**
  the quality or state of being impermeable; resisting penetration by water or plant roots. Impervious ground cover like concrete and asphalt affects quantity and quality of runoff.
- **import**
  water piped or channelled into an area.
- **imported water**
  water brought into an area from a distant source, such as from one part of a state to another via an aqueduct.
- **impoundment**
  a body of water such as a pond, confined by a dam, dike, floodgate or other barrier. It is used to collect and store water for future use.

- **increaser**
  a range plant species that is present in a range ecosystem prior to grazing and that increases in numbers or coverage as a result of grazing.
- **indicator organisms**
  micro-organisms, such as coliforms, whose presence is indicative of pollution or of more harmful micro-organism.
- **indicator tests**
  tests for a specific contaminant, group of contaminants, or constituent which signals the presence of something else (ex., coliforms indicate the presence of pathogenic bacteria).
- **indigenous plants**
  plants native to an area and biome. Also referred to as Native Plants. See **naturalised plants**.
- **induction motor (AC)**
  the type of electric motor used in conventional AC water pumps. It requires a high surge of current to start and a stable voltage supply, making it relatively expensive to run by solar power.

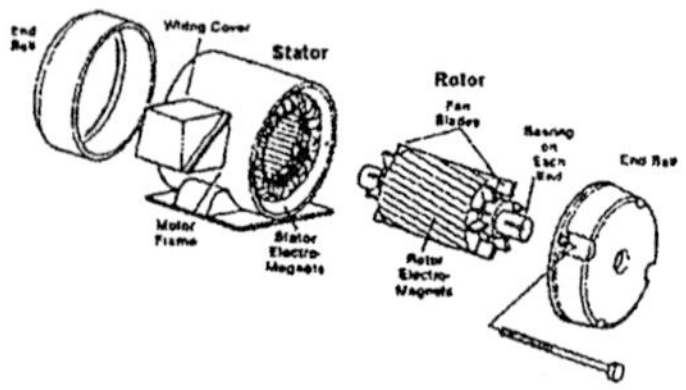

- **industrial water use**
  water used for industrial purposes in such industries as steel, chemical, paper, and petroleum refining. Nationally, water for industrial uses comes mainly (80%) from self-supplied sources, such as a local well or withdrawal points in a river, but some water comes from public-supplied sources, such as the county/city water department.

- **infiltration**
  see also **intake rate**. Process of water movement through the soil surface into the soil matrix. The act of water entering the soil profile.

- **infiltration capacity**
  the maximum rate at which the soil, when in a given condition, can absorb falling rain or melting snow.

- **infiltration index**
  an average rate of infiltration, in inches per hour, equal to the average rate of rainfall such that the volume of rain fall at greater rates equals the total direct runoff.

- **infiltration rate**
  the rate of water entry into the soil expressed as a depth of water per unit of time in inches per hour or feet per day. The infiltration rate changes with time during irrigation.

- **infiltration**
  the flow of a fluid into a substance through pores or small openings. It connotes flow into a substance in contradistinction to the word percolation, which connotes flow through a porous substance.

- **infiltrometer**
  device used to measure the infiltration rate / intake rate of water into soil.

  ring infiltrometer
  consists of metal rings that are inserted (driven) into the soil surface and filled with water. The rate at which water enters the soil is observed.

  sprinkler infiltrometer
  consists of a sprinkler head(s) that applies water to the soil

surface at a range of rates of less-than to greater than soil infiltration rates. Maximum infiltration rates are observed and recorded.

flowing infiltrometer
consists of an inlet device to apply a specific flow rate to a furrow and a collection sump with a pump to return tail water to the inlet device. Water infiltrated by the soil in the test section (typically 10 meters) is replaced with water from a reservoir to keep the flow rate constant. The rate of water infiltrated verses time is also plotted. An equation (typically for a curvilinear line) then represents the intake characteristics for that particular soil condition.

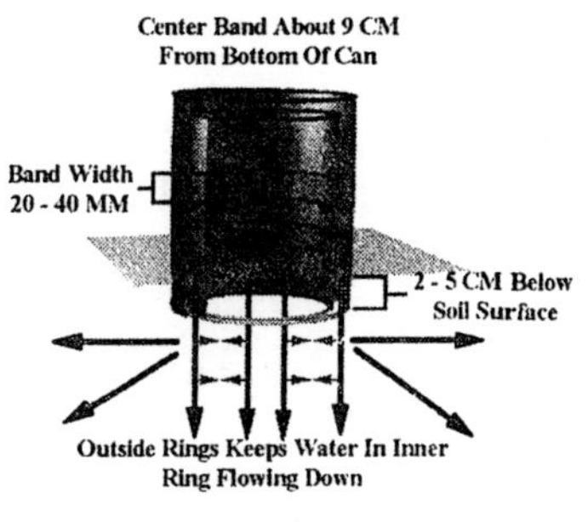

- **injection well**
refers to a well constructed for the purpose of injecting treated wastewater directly into the ground. Wastewater is generally forced (pumped) into the well for dispersal or storage into a designated aquifer. Injection wells are generally drilled into aquifers that don't deliver drinking water, unused aquifers, or below freshwater levels.

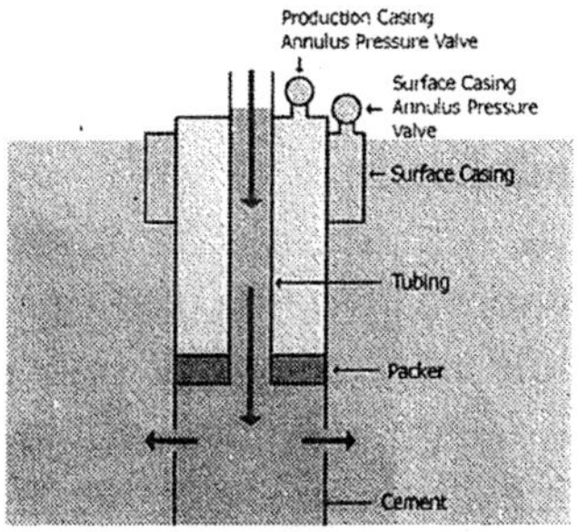

- **inland freshwater wetlands**
swamps, marshes, and bogs found inland beyond the coastal saltwater wetlands.

- **inoculum**
the initial organism or organisms that establish a new colony or population; breeding stock. The term is commonly applied to the application of small amounts of micro-organisms to establish new populations in soils or composts, for example, **Rhizobium**

bacteria are commonly applied as an inoculum to legume seeds at planting time.

- **inrush current**
  (amps) current necessary to initially open the solenoid valve.

- **insolation**
  the conversion, at the earth's surface, of short wave solar energy into long wave heat energy.

- **installation**
  the unit should be installed when soil water is at or near field capacity. The pan unit is designed to be level in the field. Under excessive slope conditions, care must be taken to ensure that the upper edge of the pan is level. The operation of the float mechanism and the position of the float depends on meeting that constraint. It is also recommended that the installation occur after plant stand establishment or when the crop is at least a couple of weeks old (roots and canopy are beginning to be established) and the soil water content is at or near field capacity (following a rain event).

- **instream flows**
  water flows for uses within a defined stream channel e.g., flows intended for fish and wildlife.

- **instream use**
  use of water that does not require withdrawal or diversion from its natural watercourse; for example, the use of water for navigation, recreation, and support of fish and wildlife.

- **insulated**
  a stream or reach of a stream that neither contributes water to the zone of saturation nor receives water from it. It is separated from the zones of saturation by an impermeable bed.

- **intake family**
  grouping of intake characteristics into families based on field infiltrometer tests on many soils.

- **intake rate (basic)**
  rate at which water percolates into the soil after infiltration has decreased to a low and nearly constant value.

- **intake rate (of soil)**
  rate that (irrigation) water enters the soil at the surface.

- **intake, initial**
depth (rate) of water absorbed by a soil during the period of rapid or comparatively rapid intake following initial application.

- **intangible**
a good, service, or effect of an action that cannot be assigned monetary value.

- **integrated pest management**
pest control using an array of complementary approaches including natural predators, parasites, pest-resistant varieties, pesticides, and other biological and environmental control practices.

- **interbasin transfer**
a movement of water from one drainage basin to another, such as from the east side of the Rocky Mountains into the west-flowing Colorado River.

- **interception**
the process and the amount of rain or snow stored on leaves and branches and eventually evaporated back to the air. Interception equals the precipitation on the vegetation minus stem flow and through fall.

- **intercropping**
planting more than one crop in a field using a regular pattern that interleaves each crop in some pattern. A form of polyculture.

- **intermediate disturbance hypothesis**
the theory that diversity and productivity in natural ecosystems are highest when moderate disturbance occurs periodically but not too frequently.

- **intermittent or seasonal**
a stream which flows only at certain times of the year when it receives water from springs, rainfall, or from surface sources such as melting snow. Perennial - A stream which flows continuously.

- **intermittent stream**
a stream which has a period of zero flow for at least one week during most years. Where flow records are available, a stream with a flow of less than 0.1 cubic feet per second is considered intermittent.

- **internal manual bleed**
feature which allows an automatic valve to be opened

manually (without controller) by releasing water from above the diaphragm to the downstream side of the valve. Useful during installation, system start-up and maintenance operations when it is undesirable for water to escape into the valve box.

- **internal waters**
  waters under the exclusive control of a coastal nation, including bays, estuaries, and rivers.

- **interspecific competition**
  competition for resources among individuals of different species.

- **interstate water**
  according to law, interstate waters are defined as
  1. rivers, lakes and other waters that flow across or form a part of state or international boundaries;
  2. waters of the Great Lakes;
  3. coastal waters whose scope has been defined to include ocean waters seaward to the territorial limits and waters along the coastline (including inland streams) influenced by the tide.

- **interstices**
  the void or empty portion of rock or soil occupied by air or water.

- **intraspecific competition**
  competition for resources among individuals of the same species.

- **introduced plant**
  a plant imported from another region. Survival depends on a similar climate or the ability to adapt to the new climate.

- **invader**
  a range plant species not present in a given area before grassing but entering the area as a result of the ecological changes caused by grazing.

- **IPM**
  see integrated pest management

- **IPS**
  see **iron pipe size.**

- **iron pipe size**
  one of several systems which specify pipe dimensions.

- **irrecoverable water loss**
  water loss that becomes unavailable for reuse through evaporation, phreatophytic

transpiration, or ground-water recharge that is not economically recoverable.

- **irrigable area**
(acres) area capable of being irrigated, principally based on availability of water, suitable soils and topography of land.

- **irrigated acreage**
irrigable acreage actually irrigated in any one year. It includes irrigated cropland harvested, irrigated pasture, cropland planted but not harvested, and the acreage in irrigation rotation used for soil building crops.

- **irrigated area**
the gross farm area upon which water is artificially applied for the production of crops, with no reduction for access roads, canals or farm buildings.

- **irrigation**
the application of water to crops by artificial means. Purposes for irrigating may include, but are not limited to, supplying evapotranspiration needs, leaching of salts and environmental control.

- **irrigation audit**
procedure to collect and present information concerning the uniformity of application, precipitation rate, and general condition of an irrigation system and its components.

- **irrigation check**
small dike or dam used in the furrow alongside an irrigation border to make the water spread evenly across the border.

- **ICUC**
irrigation consumptive use coefficient

- **irrigation design**
plan of an irrigation system with pipe sizing, head layout and valve location.

- **irrigation efficiency**
the average percent of total water pumped or delivered for use that is stored in the plants root zone.

- **irrigation frequency**
measure of the number of irrigations per unit time.

- **irrigation interval**
average time interval between the commencement of successive irrigations for a given field (or area). Time

between irrigation events. Usually considered the maximum allowable time between irrigations during the peak ET period.

- **irrigation period**
  time that it takes to apply one irrigation to a given design area during the peak consumptive-use period of the crop being irrigated.

- **irrigation requirement**
  the quantity of water, exclusive of precipitation, that is required for crop production. It includes surface evaporation and other economically unavoidable wastes.

- **irrigation return flow**
  water which is not consumptively used by plants and returns to a surface or ground water supply. Under conditions of water right litigation, the definition may be restricted to measurable water returning to the stream from which it was diverted.

- **irrigation sagacity**
  ratio of volume of irrigation water beneficially or reasonably used to the total volume of irrigation water that has left the region, both in a specified period of time.

- **irrigation schedule**
  procedure of establishing and implementing the time and amount of irrigation water to apply. Determining when to irrigate and how much water to apply, based upon measurements or estimates of soil moisture or crop water used by a plant. Set of specifications identifying times to turn on and off water to various zones of an irrigation system.

- **irrigation set**
  (acres) area irrigated at one time within a field.

- **irrigation slope (%, ft / 100 ft)**
  elevation difference along the direction of irrigation. Sometimes called irrigation grade.

- **irrigation system**
  physical components (pumps, pipelines, valves, nozzles, ditches, gates, siphon tubes, turnout structures) and management used to apply irrigation water by an irrigation method.

• **irrigation uniformity**
a measure of the spatial variability of applied or infiltrated water over the field.

• **irrigation water**
water which is applied to assist crops in areas or during times where rainfall is inadequate.

• **irrigation water use**
water application on lands to assist in the growing of crops and pastures or to maintain vegetative growth in recreational lands, such as parks and golf courses.

• **irrigation(water) requirement**
depth of water, exclusive of effective precipitation, stored soil , moisture or ground water, that is required for meeting crop evapotranspiration for crop production and other related uses. Such uses may include water required for leaching, frost protection, cooling and chemigation. Difference between potential evapotranspiration and effective precipitation. Quantity of water needed by the landscape to satisfy the evaporation, transpiration and other uses of the water in the soil.

• **irrigation, supplemental**
see **supplemental irrigation**

• **isohyet**
line that connects points of equal rainfall. See **isohyetal line**

• **isohyetal line (isohyet)**
a line drawn on a map or chart joining points that receive the same amount of precipitation.

• **isolation valve**
a valve used for isolating all or part of the irrigation system for repairs, maintenance or winter shut-down (winterisation). Common types of isolation valves are the ball valve, butterfly valve, gate valve and white bread.

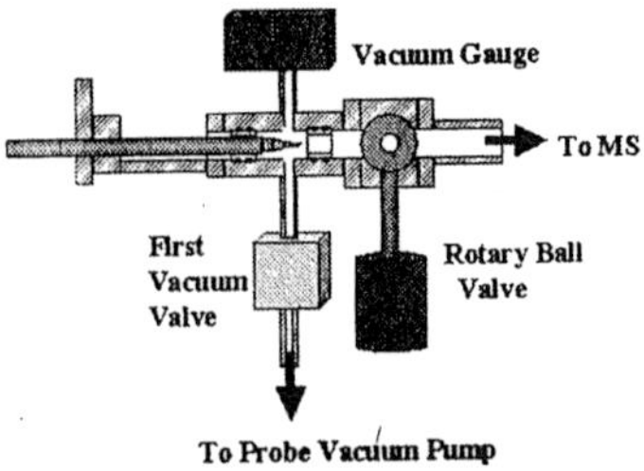

- **isotherm**
  line that connects points of equal temperature.

- **jet pump**
  a surface-mounted centrifugal pump that uses an 'ejector' (venturi) device to augment its suction capacity. In a 'deep well jet pump', the ejector is down in the well, to assist the pump in overcoming the limitations of suction. (Some water is diverted back down the well, causing an increase in energy use.)

- **jet stream**
  a long narrow meandering current of high-speed winds near the tropopause blowing from a generally westerly direction and often exceeding a speed of 250 miles per hour.

- **jetteau**
  a jet of water.

- **jetter**
  one (as a geyser) that sends out a jet.

- **jockey pump**
  see **pump.**

- **kalema**
  a violent surf that occurs on the coast of the Guinea region, West Africa.

- **kame**
  a short ridge, hill, or mound of stratified drift deposited by glacial meltwater.

- **kame terrace**
  a terrace of stratified sand and gravel deposited by streams between a glacier and an adjacent valley wall.

- **kilowatthour (kWh)**
  a power demand of 1,000 watts for one hour. Power company utility rates are typically expressed in rupees per kilowatt-hour.

- **kinematic viscosity (ft^2/s)**
  measure of the resistance of a liquid to shear forces.

- **k-y jelly**
  a clear, water soluble, non-toxic lubricant used for lubricating O-rings in water filters. Also sometimes used to lubricate threads and ease insertion of drip tube into fittings. Because it is water soluble it doesn't gum up emitters, filters, or sprinkler mechanisms.

- **labour theory of value**
  an approach to valuing commodities based on the amount of human labour required to produce them.

- **laboratory water**
  purified water used in the laboratory as a basis for making up solutions or making dilutions. Water devoid of interfering substances.

- **lag**
  variously defined as time from beginning (or centre of mass) of rainfall to peak (or centre of mass) of runoff.

- **lag time (flood irrigation)**
  (h, min)Period between the time that the irrigation stream is turned off at the upper end of an irrigated area and the time that water disappears from the surface at the point or points of application.

- **lagoon**
  a shallow pond where sunlight, bacterial action, and oxygen work to purify wastewater. Lagoons are typically used for the storage of wastewaters, sludges, liquid wastes or spent nuclear fuel.

- **lake**
  an inland body of water, usually fresh water, formed by glaciers, river drainage etc. Usually larger than a pool or pond.

- **land classification**
reclamation's systematic placing of lands into classes based on their suitability for sustained irrigated farming. Land classes are defined by productivity, with Class 1 being the most productive. For other classes, the equivalent acreage to Class 1 for the same productivity is defined (Class 1 equivalency).

- **land retirement**
permanent removal of land from agricultural production.

- **land utilisation type (LUT)**
a kind of land use described or defined in a degree of detail greater than that of a major kind of land use. In the context of irrigated agriculture, a land utilisation type refers to a crop, crop combination or cropping system with specified irrigation and management methods in a defined technical and socio-economic setting. In the context of rainfed agriculture, a land utilisation type refers to a crop, crop combination or cropping system with a specified technical and socio-economic setting. A land utilisation type in forestry consists of a technical specifications in a given physical, economic and social setting.

- **landrace**
a locally-adapted strain of a species bred through traditional methods of directed selection. Also called farmer varieties.

- **landscape ecology**
the study of environmental factors and interactions at a scale that encompasses more than one ecosystem at a time.

- **landscape impoundment**
body of reclaimed water which is used for aesthetic enjoyment or which otherwise serves a function not intended to include contact recreation.

- **laser soaker line**
1/4" tubing with preset emission holes equally spaced along its entire length, providing a continuous wetted pattern.

- **lateral**
the name given to the pipes which go from the control valves to the sprinklers or drip emitter tubes.

- **lateral movement**
  refers to the movement of water in the soil in a horizontal direction.

- **laterals**
  the actual drip lines, either polyethylene hose with emitters attached at desired intervals or laser soaker line.

- **lath box**
  wooden box that is placed in a ditch bank to transfer water from an irrigation ditch to the field to be irrigated (preferred term is spile).

- **leaching**
  extraction or flushing out of dissolved or suspended materials from the soil, solid waste, or another medium by water or other liquids as they percolate down through the medium to groundwater.

- **leaching fraction (LF)(-0)**
  ratio of the depth of subsurface drainage water (deep percolation) to the depth of infiltrated irrigation water see **leaching requirement**.

- **leaching requirement**
  quantity of irrigation water required for transporting salts through the soil profile to maintain a favourable salt balance in the root zone for plant development.

- **length of run**
  distance water must flow in furrows or borders over the surface of a field from the head to the end of the field.

- **lentic system**
  a nonflowing or standing body of fresh water, such as a lake or pond. Compare **lotic system**.

- **lentic waters**
  ponds or lakes (standing water).

- **LEPA**
  acronym for low energy precision application.

- **levee**
  a natural or man-made earthen obstruction along the edge of a stream, lake, or river. Usually used to restrain the flow of water out of a river bank.

- **life zone**
  attitudinal or latitudinal biotic regions with distinctive flora and fauna, characterised by plant and animal biomes.

- **light compensation point**
  the level of light intensity needed for a plant to produce an amount of photosynthate equal to the amount it uses for respiration.

- **light reactions**
  the components of photosynthesis in which light energy is converted into chemical energy in the form of ATP and NADPH.

Light reactions of photosynthesis

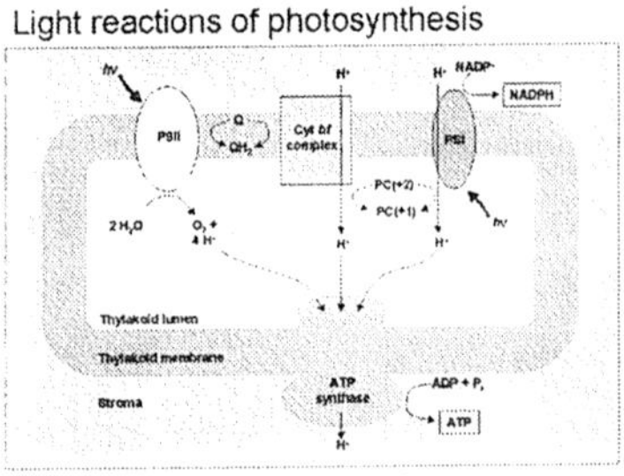

- **light soil**
  a term sometimes used to describe a soil with large particles loosely packed together.

- **limited irrigation**
  management of irrigation applications to apply less than enough water to satisfy the soil water deficiency in the entire root zone. Sometimes called ' deficit' or 'stress irrigation'.

- **limiting factor**
  factor such as temperature, light, water, or a chemical that limits the existence, growth, abundance, or distribution of an organism.

- **limiting nutrient**
  a nutrient not present in the soil in sufficient quantity to support optimal plant growth.

- **limits to growth**
  a world model developed in the 1970s by a group called the Club of Rome, it predicted resource scarcity if world population and resource use growth continued.

- **limnology**
  that branch of hydrology pertaining to the study of lakes, ponds and streams.

- **line source**
  continuous source of water emitted along a line.

- **linear current booster**
  a 'matching' device to convert the (usually) higher DC voltage from the solar panels into a somewhat lower voltage but higher current for the pump. See **pump controller** and **maximum power point tracker**.

- **line-source emitter**
water is discharged from closely spaced perforations, emitters or a porous wall along the tubing.

- **lining**
protective covering over the perimeter of a conduit, reservoir or channel to prevent seepage losses, to withstand pressure or to resist erosion.

- **liquid**
a state of matter, neither gas nor solid, that flows and takes the shape of its container.

- **litres per minute**
a metric measurement of water flow used worldwide.

- **littoral zone**
area on or near the shore of a body of water.

- **livestock water use**
water used for livestock watering, feed lots, dairy operations, fish farming, and other on-farm needs.

- **living mulch**
a cover crop that is interplanted with the primary crop(s) during the growing season.

- **local runoff**
water running off a local area, such as rainfall draining into a nearby creek.

- **long path emitter**
employs a long capillary-sized tube or channel to dissipate pressure.

- **long-period variations**
secular when a cycle or a change in trend is completed within a century; climatic when the period of change runs through centuries or a few millennia, geologic when the period runs into geological time.

- **looped circuit**
piping system, usually a main line, that closes back on itself in a loop, thus providing water to any location from two routes.

- **losing**
a stream or reach of a stream that contributes water to the zone of saturation. An influent stream.

- **lotic system**
a flowing body of fresh water, such as a river or stream. Compare **lentic system**.

- **lotic waters**
  flowing waters, as in streams and rivers.

- **low energy precision application(LEPA)**
  a water, soil and plant management regime where precision down-in-crop applications of water are made on the soil surface at the point of use. Application devices are located in the crop canopy on drop tubes mounted on low pressure centre pivot and linear move sprinkler irrigation systems.

- **low head drainage**
  condition in which water drains partially or completely out of the lateral line through the sprinkler head after each irrigation cycle is completed.

- **low pressure in canopy (LPIC)**
  low-pressure in-canopy system that may or may not include a complete water, soil and plant management regime as required in LEPA. Application devices are located in the crop canopy with drop tubes mounted on low-pressure centre pivot and linear move sprinkler irrigation systems

- **low-flow frequency curve**
  a graph showing the magnitude and frequency of minimum flows for a period of given length. Frequency is usually expressed as the average interval, in years, between recurrences of an annual minimum flow equal to or less than that shown by the magnitude scale.

- **lysimeter**
  structure containing a mass of soil, and designed to permit the measurement of water draining through the soil.

- **MAD**
  (management allowed deficit or depletion) is the percentage of the available water in the soil that will be allowed to be removed by plants before irrigation is required. MAD is primarily a function of the type of crop being grown. For high value crops, MAD may be 40 percent or less. For lower value or large orchard crops (such as trees) MAD may be 60 percent. A MAD of 50 percent is reasonable for most row crops.

- **main (mainline)**
  water delivery pipelines that supply water from the control station to the manifolds. Pipe usually under constant pressure which supplies water from the point of connection to the control valves (or valve-in-head sprinklers).

- **male adapter**
  the fitting used to adapt from solvent weld PVC to a male threaded end. When connecting to metal threads male adapters should be used, so that the plastic male threads screw into the metal female threads.

- **management allowable (allowed) depletion(deficit)**
  see similar term, **maximum allowable deficiency**.
  desired soil moisture deficit at the time of irrigation. Portion of available water that is scheduled to be used prior to the next irrigation. Planned soil moisture deficit at the time of irrigation.

- **manifold**
  pipeline that supplies water to the laterals. Closely linked series of mainline piping supplying water to valves or laterals.

- **manufacturer's coefficient of variation (C_v)(-)**
  measure of the variability of discharge of a random sample of a given make, model, and size of micro-irrigation emitter, as produced by the manufacturer and before any field operation or ageing has taken place; equal to the ratio of the standard deviation of the discharge of the emitters to the mean discharge of the emitters.

- **manure**
  animal excrement in varying stages of decomposition. Manure should be aged for landscape use. Fresh manure is high in nitrogen and can have a high salt content, which can harm plant roots.

- **marsh**
  a frequently or continually inundated wetland characterised by emergent herbaceous vegetation adapted to saturated soil conditions.

- **mass curve**
  a graph of the cumulative values of a hydrologic quantity (such as precipitation or runoff), generally as ordinate, plotted against time or date as abscissa.

- **matched precipitation rate**
system or zone in which all the heads have similar precipitation rates is said to have matched precipitation rates.

- **matric potential**
dynamic soil property, will be near zero for a saturated soil. Matric potential results from capillary and adsorption forces. This potential was formerly called capillary potential or capillary water.

- **Maximum Allowable Deficiency**
See similar term, **management allowed depletion**. term used to estimate the amount of water that can be used without adversely affecting the plant and is defined as the ratio of readily available water to available water.

- **maximum application rate**
(in./h) maximum discharge, in at which sprinklers can apply water without causing significant translocation.

- **maximum power point tracker**
an electronic device that acts as a 'transmission' between the PV panels and the pump. Provides the maximum power possible out of the solar array. While an array tracker that follows the sun provides higher efficiency and power in the summer, a MPPT provides the highest gains in winter and/or cold weather, due to the higher PV outputs at colder cell temperatures. Where maximum efficiency is needed, both may be used.

- **maximum probable flood**
see **flood, maximum probable**

- **maximum sustainable yield**
the largest average harvest of a species that can be indefinitely sustained under existing environmental conditions.

- **meadow**
in geologic terms, a meadow is a lake that has filled with silt. In general landscape terms, a meadow has come to refer to a level, open space with mixed grasses and perennial flowers.

- **meander**
the winding of a stream channel

- **meander amplitude**
distance between points of maximum curvature of suc-

cessive meanders of opposite phase in a direction normal to the general course of the meander belt, measured between centrelines of channels.

- **meander belt**
  area between lines drawn tangential to the extreme limits of fully developed meanders.
- **meander breadth**
  the distance between the lines used to define the meander belt.
- **meander length**
  distance in the general course of the meanders between corresponding points of successive meanders of the same phase.
  Twice the distance between successive points of inflection of the meander wave.
- **median drop size**
  diameter where half the sprinkler's water volume falls in drops smaller, and half falls in drops larger than the median size. Drop size where 50% of the water volume occurs in drops greater than this size.
- **median streamflow**
  the rate of discharge of a stream for which there are equal numbers of greater and lesser flow occurrences during a specified period.
- **melting**
  the changing of a solid into a liquid.
- **meltwater**
  water that comes from the melting ice of a glacier or a snowbank.
- **meromictic lake**
  a lake in which some water remains partly or wholly unmixed with the main water mass at circulation periods is said to be meromictic. The process leading to a meromictic state is termed meromixis the perennially stagnant deep layer of a meromictic lake is called the monimolimnion. The part of a meromictic lake in which free circulation can occur is called the mixolimnion. The boundary between the monimolimnion and the mixolimnion is called thechemocline.
- **mesosphere**
  layer of the atmosphere between 30 and 50 mi (50 and 80

km) in altitude, characterised by decreasing temperatures with increasing altitude.

- **meteoric water**
  new water derived from the atmosphere.
- **method blank**
  laboratory grade water taken through the entire analytical procedure to determine if samples are being accidentally contaminated by chemicals in the lab
- **mg**
  million gallons.
- **mg/l**
  milligrams per litre.
- **mgd**
  million gallons per day.
- **mgy**
  million gallons per year.
- **MHT**
  Male Hose Thread (as found on the end of a garden hose).
- **micro-irrigation**
  the application of water directly to, or very near to the soil surface in drops, small streams, or sprays.
- **micro-climate**
  1. atmospheric conditions within or near a crop canopy.
  2. an area influenced by natural or manmade features that alter the climatic conditions from the general regional climate. The climate alterations may include but are not limited to changes in temperature range, wind exposure, reflected light, shelter belts and precipitation rates.
- **micrograms per litre - μg/l**
  micrograms per litre of water. One thousands micrograms per litre is equivalent to 1 milligram per litre. This measure is equivalent to parts per billion (ppb)
- **migration**
  the movement of oil, gas, contaminants, water or other liquids through porous and permeable rock.
- **mildew**
  a group of diseases caused by fungi and characterised by a white powder-like substance on the leaves. Often caused by poor air circulation around the plant or plants.
- **milligrams per litre (mg/l)**
  a unit of the concentration of a constituent in water or wastewater. It represents 0.001 gram of a constituent

in 1 litre of water. It is approximately equal to one part per million (PPM).

- **mined-land reclamation**
  the return of land disturbed by mining to a more productive condition, usually a use similar to that existing before mining took place.

- **minimum streamflow**
  the specific amount of water reserved to support aquatic life, to minimise pollution or for recreation. It is subject to the priority system and does not affect water rights established prior to its institution.

- **minimum tillage**
  a soil and water conservation technique that leaves the crop residue or stubble on the surface rather than ploughing it under to minimise the number of times a field is tilled. Weeds are controlled by herbicides.

- **mining water use**
  water use during quarrying rocks and extracting minerals from the land.

- **mini-spray**
  a variety of low volume overhead sprays which cover larger areas than an emitter.

- **MIPT**
  acronym for male iron pipe thread.

- **mist irrigation**
  method of micro-irrigation in which water is applied in very small droplets.

- **mixed cropping**
  an agricultural system in which several different crops are grown in close proximity, in a rotation system, or both.

- **mobile irrigation laboratory**
  a vehicle furnished with irrigation evaluation equipment which is used to carry out on-site evaluations of irrigation systems and to provide recommendations on improving irrigation efficiency.

- **moisture**
  water diffused in the atmosphere or the ground.

- **moisture deficit, soil moisture depletion**
  difference between actual soil moisture and soil moisture held in the soil at field capacity.

- **moisture equivalent**
  the ratio of
  the weight of water which the soil, after saturation,

will retain against a centrifugal force 1,000 times the force of gravity, to the weight of the soil when dry. The ratio is stated as a percentage.

- **moisture meter**
  device that monitors or measures soil water content or tension.

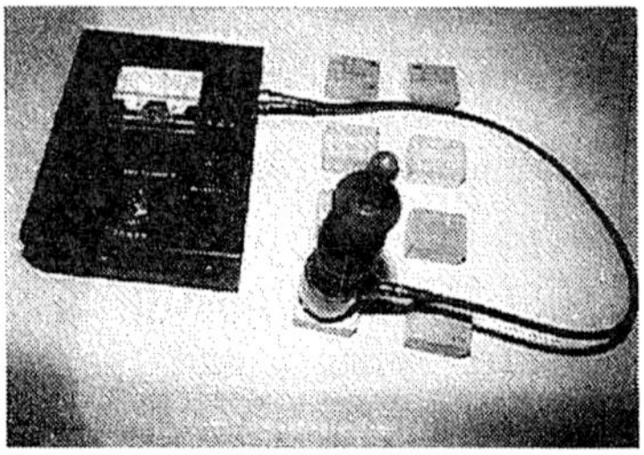

- **moisture sensor**
  instrument which monitors soil water content. Device that monitors or measures soil water content of tension.

- **monoculture**
  an agricultural system in which a single crop is grown repeatedly over a large area.

- **MPT**
  male nominal pipe threads.

- **mudflow**
  a well-mixed mass of water and alluvium which, because of its high viscosity and low fluidity as compared with water, moves at a much slower rate, usually piling up and spreading over the fan like a sheet of wet mortar or concrete.

- **mulch**
  a protective covering of various substances, especially organic, placed on the ground around plants to reduce weed growth and evaporation of moisture from the soil surface, and to maintain even temperatures around plant roots.

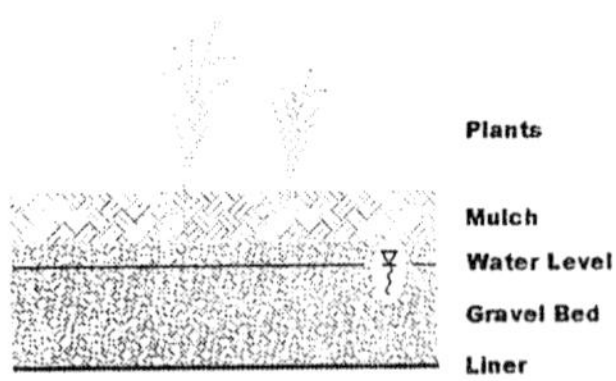

- **multi-outlet emitter**
  supplies water to two or more points through small diameter auxiliary tubing.

- **multiple use**
  the use of lands for as many different purposes as possible in order to gain maximum benefit from them.

- **multi-stage centrifugal**
  a centrifugal pump with more than one impeller and chamber, stacked in a sequence to produce higher pressure. Conventional AC deep well submersible pumps and higher power solar submersibles work this way.

- **municipal sewage**
  sewage from a community which may be composed of domestic sewage, industrial wastes or both.

- **municipal solid waste**
  mixed solid waste derived primarily from residential and commercial sources.

- **municipal water system**
  a water system that has at least five service connections or which regularly serves 25 individuals for 60 days; also called a public water system

- **natural capital**
  the stock of natural resources that provide or can be used to make beneficial things.

- **natural flow**
  the rate of water movement past a specified point on a natural stream. The flow comes from a drainage area in which there has been no stream diversion caused by storage, import, export, return flow, or change in consumptive use caused by man-controlled modifications to land use. Natural flow rarely occurs in a developed country.

- **natural resource**
  something that is useful to humans and exists independent of human activity.

- **natural resource accounting**
  the inclusion of the full and often hidden costs of damages to natural resources in traditional benefit/cost analysis.

- **naturalistic planting**
  refers generally to the arrangement of plants and plant selection. Naturalistic plantings are characterised by groupings and forms reflecting how plants would grow in an undisturbed environment.

- **naturalised plants**
  plants introduced from another locality, but thoroughly established in the area. Many

plants that are thought to be native are in fact naturalised. Many weeds are naturalised plants..

- **nephelometric turbidity unit (NTU)**
  unit of measure for the turbidity of water. Essentially, a measure of the cloudiness of water as measured by a nephelometer. Turbidity is based on the amount of light that is reflected off particles in the water.

- **net (irrigation) water requirement**
  same as net irrigation requirement.

- **net positive suction head (NPSH)**
  head that causes liquid to flow through the suction piping and enter the eye of the pump impeller. Net positive suction head available. Pressure head that is supplied (is available) to the eye of an impeller in a pump based on system characteristics. Net positive suction head required. Minimum pressure head required at the eye of an impeller in a pump to prevent cavitation.

- **net precipitation rate**
  see **precipitation rate**

- **net primary production**
  the net amount of biomass created by plants in an ecosystem once the respiration by those plants is deducted.

- **neutral stuff**
  something that exists but at present meets no known human material or non-material needs.

- **neutron probe**
  an instrument used to estimate soil moisture. Relates the rate of attenuation in pulsed neutron emissions to soil water content.

- **nitrogen**
  a plant nutrient that can cause an overabundance of bacteria and algae when high amounts are present, leading to a depletion of oxygen and fish kills. Several forms occur in water, including ammonia, nitrate, nitrite or elemental nitrogen. High levels of nitrogen in water are usually caused by agricultural runoff or improperly operating wastewater treatment plants.

- **nominal**
named size which is usually not the actual dimensions of the product. i.e. a half inch schedule 40 pipe is not ½ inch ID or OD .

- **non-alkali soil**
acid soil with a pH below seven. See **acid/alkaline soil**.

- **nonconsumptive use**
using water in a way that does not reduce the supply. Examples include hunting, fishing, boating, water-skiing, swimming and some power production. Compare **consumptive** use.

- **noncontact recreation**
recreational pursuits not involving a significant risk of water ingestion, including fishing, commercial and recreational boating, and limited body contact incidental to shoreline activity. Compare **contact recreation**.

- **nonpoint source**
a pollution source that is diffuse, such as urban runoff. Non-renewable or stock resources: Resources that exist in finite quantity and are not replaced in nature.

- **non-point source (NPS) pollution**
pollution discharged over a wide land area, not from one specific location. These are forms of diffuse pollution caused by sediment, nutrients, organic and toxic substances originating from land-use activities, which are carried to lakes and streams by surface runoff. Non-point source pollution is contamination that occurs when rainwater, snowmelt, or irrigation washes off ploughed fields, city streets or suburban backyards. As this runoff moves across the land surface, it picks up soil particles and pollutants, such as nutrients and pesticides.

- **nonporous**
something which does not allow water to pass through it.

- **nonpotable**
not suitable for drinking.

- **non-saline sodic soil**
soil containing soluble salts that provide an electrical conductivity of saturation extract (ECe) less than 4.0 mmhos/cm and an exchangeable sodium per-

centage (ESP) greater than 15. Commonly called black alkali or slick spots.

- **nonthreshold pollutant**
substance or condition harmful to a particular organism at any level or concentration.

- **nozzle**
the part of a sprinkler that the water comes out of. Usually a very carefully engineered part to assure a good spray pattern. In most cases, the nozzle is removable so that it can be easily cleaned or replaced. With plastic nozzles replacement is generally preferred over cleaning as small scratches in the plastic can cause big problems with water distribution uniformity.

- **NTU**
nephlometric turbidity units.

- **number of outlets**
(N_0)term used to describe the number of outlets in a lateral.

- **nutrient**
as a pollutant, any element or compound, such as phosphorous or nitrogen, that fuels abnormally high organic growth in aquatic ecosystems.

- **off-stream use**
use of water that requires that it be removed or withdrawn from surface or groundwater.

- **oligotrophic**
describes lakes that are relatively deep and nutrient poor.

- **on centre**
industry-standardised term referring to the distance or spacing between the centremost point of each plant.

- **on-farm**
activities (especially growing crops and applying irrigation water) that occur within the legal boundaries of private property.

- **on-farm irrigation efficiency**
the ratio of the volume of water used for consumptive use and leaching requirements in cropped areas to the volume of water delivered to a farm (applied water).

- **open discharge**
the filling of a water vessel that is not sealed to hold pressure. Examples: storage

(holding) tank, pond, flood irrigation.

- **open system**
system in which energy and matter are exchanged between the system and its environment, for example, a living organism.

- **operating pressure**
the pressure at which a device or irrigation system is designed to operate. Can mean just about anything depending on usage. There can be 'optimum operating pressure' ,'minimum operating pressure', 'maximum operating pressure' and 'operating pressure range'.

- **operational losses**
losses of water resulting from evaporation, seepage and spills.

- **operational waste**
water that is lost or otherwise discarded from an irrigation system after having been diverted into it as part of normal operations.

- **opportunity time**
time that water inundates the soil surface with opportunity to infiltrate.

- **organic chemicals**
chemicals containing carbon.

- **organic material**
any material of organic origin, such as ground bark, sphagnum peat moss, compost, aged manure, which can be mixed into the soil to improve its condition.
All are based upon carbon compounds.

- **organics**
being composed of or containing matter of plant and animal origin.

- **orifice**
opening with a closed perimeter through which water flows. Certain shapes of orifices are calibrated for use in measuring flow rates. Opening in system component such as pipe, tubing or nozzle.

- **orifice emitter**
employs a series of orifices to dissipate pressure.

- **orogeny**
period of mountain-building.

- **orographic precipitation**
rainfall that occurs as a result of warm, humid air being forced to rise by topo-

graphic features such as mountains.

- **osmosis**
  the movement of water molecules through a thin membrane. The osmosis process occurs in our bodies and is also one method of desalinising saline water.

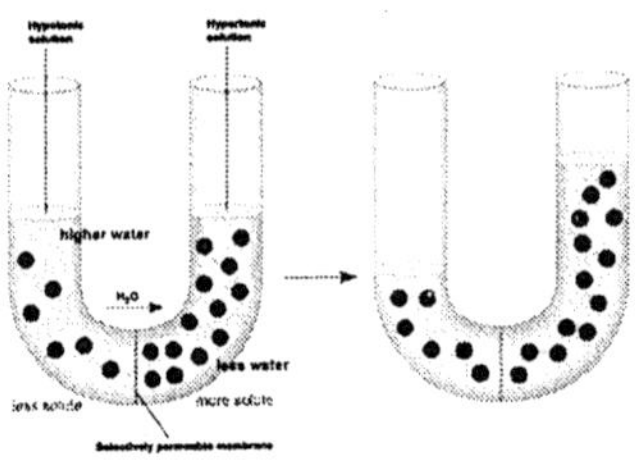

- **osmotic potential**
  potential attributable to the presence of solutes in the soil- in other words, to the soil solution.

- **outfall**
  the place where a sewer, drain or stream discharges; the outlet or structure through which reclaimed water or treated effluent is finally discharged to a receiving water body.

- **outwash**
  a deposit of sand and gravel formed by streams of meltwater flowing from a glacier.

- **oven dry**
  drying of soil samples in an over for a sufficient period of time to reach a constant weight. Refers to soil samples that have been dried in an oven at 105 °C for 24 hours

- **overdraft**
  condition that occurs in a groundwater basin when pumping exceeds the amount of replenishment over a period of years

- **overdraft or groundwater mining**
  withdrawal of groundwater in excess of the replacement rate over a long period of time.

- **overgrazing**
  grazing by a number of animals exceeding the carrying capacity of a given parcel of land.

- **overhead irrigation**
  same as sprinkler irrigation.

- **overland flow**
  the flow of rainwater or snowmelt over the land surface toward stream channels. After it enters a stream, it becomes runoff.

• **overlap**
area which is watered by two or more sprinklers.

• **oxygen demand**
the need for molecular oxygen to meet the needs of biological and chemical processes in water. Even though very little oxygen will dissolve in water, it is extremely important in biological and chemical processes.

• **oxygen demanding waste**
organic water pollutants that are usually degraded by bacteria if there is sufficient dissolved oxygen (DO) in the water.

• **ozone hole**
a semi-permanent depletion in stratospheric ozone

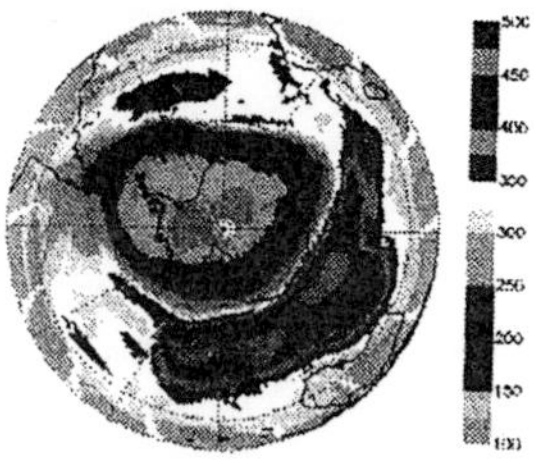

concentration over a polar region. Most prominent over the South Pole.

• **pan evaporation**
evaporative water losses from a standardised pan. Pan evaporation is sometimes used to estimate crop evapotranspiration and assist in irrigation scheduling.

• **parent material**
the mineral matter from which soil is formed.

• **Parshall flume**
a calibrated device, based on the principle of critical flow, used to measure the flow of water in open conduits. Formerly termed the Improved Venturi Flume.

• **partial-duration flood series**
a list of all flood peaks that exceed a chosen base stage or discharge, regardless of the number of peaks occurring in a year.

• **particle size**
the diameter, in millimetres, of suspended sediment or bed material. Particle-size classifications are:

1. Clay—0.00024-0.004 millimetres (mm);
2. Silt—0.004-0.062 mm;
3. Sand—0.062-2.0 mm; and
4. Gravel—2.0-64.0 mm.

- **particulate matter**
  in reference to air quality, solid or liquid particles with diameters from 0.03 to 100 microns.

- **parts per billion**
  the number of 'parts' by weight of a substance per billion parts of water. Used to measure extremely small concentrations.

- **passive solar power**
  the collection of solar energy as heat at the end-use site, without any mechanical redistribution or storage of the energy.

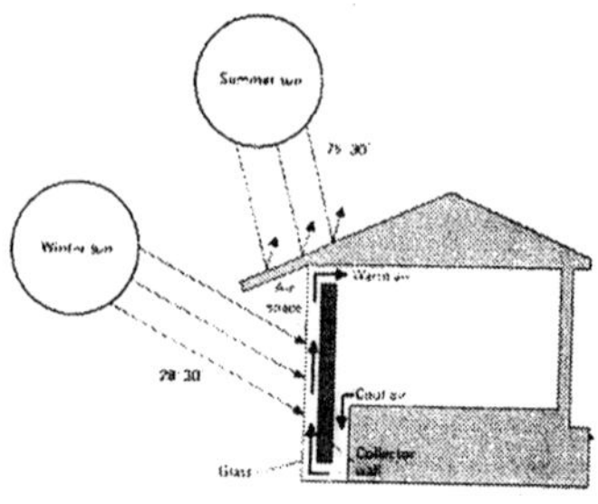

- **peak flow**
  in a wastewater treatment plant, the highest flow expected to be encountered under any operational conditions, including periods of high rainfall and prolonged periods of wet weather.

- **peak use rate**
  maximum rate at which a crop uses water.

- **peat**
  the accumulated remains of plants, found in swampy or cool, humid areas. It is the initial material from which coal may be formed, may be dried and used for fuel.

- **peat moss**
  a high-quality, organic material that is partially decomposed organic matter. Sphagnum peat is said to be the highest in quality, not to be confused with mountain peat.

- **penstock**
  a gate or sluice used in controlling the flow of water. A

tube or trough for carrying water to a water wheel, or a pipe carrying water to an electric turbine.

- **per capita use**
the average amount of water used per person during a standard time period, generally per day.

- **percentage fines**
(%)percentage of water volume falling in fine (< 1 mm in diameter) drops. Term also used relative to soil particle size.

- **perched**
a perched stream is either a losing stream or an insulated stream that is separated from the underlying ground water by a zone of aeration.

- **perched water table**
groundwater standing unprotected over a confined zone.

- **percolating waters**
waters which pass through the ground beneath the earth's surface without a definite channel. It is presumed that ground waters percolate.

- **percolation**
1. the movement of water through the openings in rock or soil.
2. the entrance of a portion of the streamflow into the channel materials to contribute to ground water replenishment.

- **percolation rate**
1. the rate at which water moves through porous media, such as soil
2. intake rate used for designing wastewater absorption systems.

- **percolation, deep**
in irrigation or farming practice, the amount of water that passes below the root zone of the crop or vegetation.

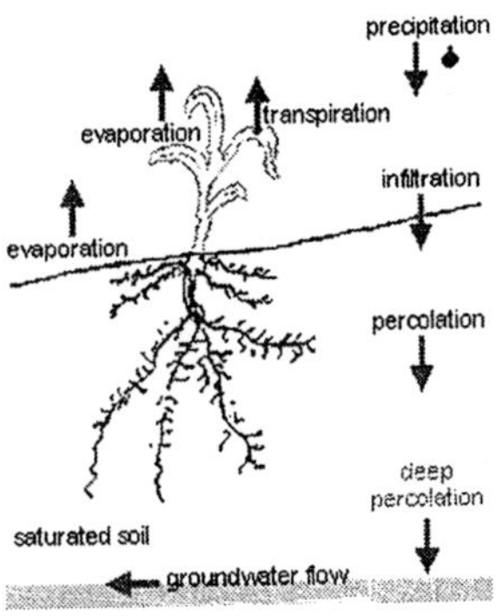

- **perennial**
a plant that lives more than two or three years. Usually

used in reference to non-woody plants.

- **perennial stream**
one that flows all year round.

- **perennial yield**
maximum quantity of water that can be annually withdrawn from a groundwater basin over a long period of time (during which water supply conditions approximate average conditions) without developing an overdraft condition.

- **perforated pipe (sprinkler)**
pipe designed to discharge water through small, multiple, closely spaced orifices or nozzles, placed in a segment of its circumference for irrigation purposes.

- **perforation of wells**
holes in the casing of wells which allow water to flow into the well.

- **perforations**
slits cut into the well casing to allow groundwater to enter. May be located at more than one level, to coincide with water-bearing strata in the earth.

- **performance curve**
graph showing the capability of a product with varying inputs i.e. the dynamic head of a pump as it varies with discharge.

- **performance-based resource estimate**
an estimate of the quantity of a mineral deposit available in the earth based primarily on the ability of prevailing technology to extract the mineral under existing and probable future economic conditions.

- **permafrost**
ground below 32°F (0°C) all year round.

- **Permanent (Wilting Point) (PWP)**
(%,in., in./ft)
moisture content, on a dry weight basis, at which plants can no longer obtain sufficient moisture from the soil to satisfy water requirements. Plants will not fully recover when water is added to the crop root zone once permanent wilting point has been experienced. Classically, 15 atmospheres (15 bars), soil moisture tension is used to estimate PWP.

Moisture content of the soil after the plant can no longer extract moisture at a sufficient rate for wilted leaves to recover overnight or when placed in a saturated environment. Also known as wilting percentage; wilting coefficient or wilting point.

- **permanent irrigation**
  irrigation having underground piping with risers and sprinklers. Preferred term is stationary sprinklers.

- **permeability**
  the ability of a water bearing material to transmit water. It is measured by the quantity of water passing through a unit cross section, in a unit time, under 100 percent hydraulic gradient.

- **permeameter**
  device for containing the soil sample and subjecting it to fluid flow in order to measure permeability or hydraulic conductivity.

- **perpetual resources**
  resources that exist in continual supply, no matter how much they are used. Solar energy is an example.

- **pet**
  (potential evapotranspiration) is the rate of water removal by the combination of plants (transpiration) and surfaces (evaporation) which is not limited by water availability.

- **petiole**
  the leaf stalk.

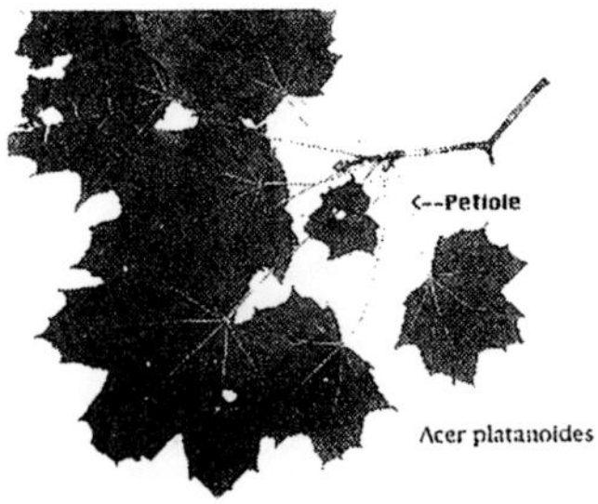

- **pH**
  a measure of the relative acidity or alkalinity of water. Water with a pH of 7 is neutral. Lower pH levels indicate increasing acidity, while pH levels higher than 7 indicate increasingly basic solutions.

- **phosphorous**
  a plant nutrient that can cause an overabundance of bacteria and algae when high amounts are present, leading to a depletion of oxygen and fish kills. High levels of phosphorous in water are usually caused by

agricultural runoff or improperly operating wastewater treatment plants. Also see **nitrogen**.

- **photosynthesis**
  the formation of carbohydrates from carbon dioxide and water, utilising light as energy.

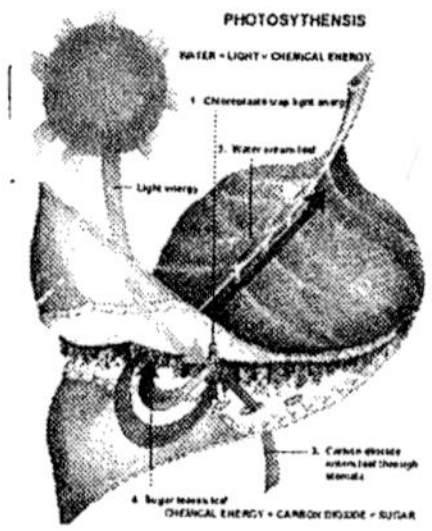

- **phreatic**
  of or relating to groundwater.

- **phreatophytes**
  plants that send their roots into or below the capillary zone to use ground water.

- **physical weathering**
  breaking down of parent rock into bits and pieces by exposure to temperature and changes and the physical action of moving ice and water, growing roots and human activities such as farming and construction.

- **piezometroc surface**
  the imaginary surface to which groundwater rises under hydrostatic pressure in wells or springs.

- **pipe dope**
  common name for commercial products used to apply to pipe fittings to assist in the appropriate fit of the threaded joints.

- **pitless adapter**
  a special pipe fitting that fits on a well casing, below ground. It allows the pipe to pass horizontally through the casing so that no pipe is exposed above ground where it could freeze. The pump may be installed and removed without further need to dig around the casing. This is done by using a 1 inch threaded pipe as a handle.

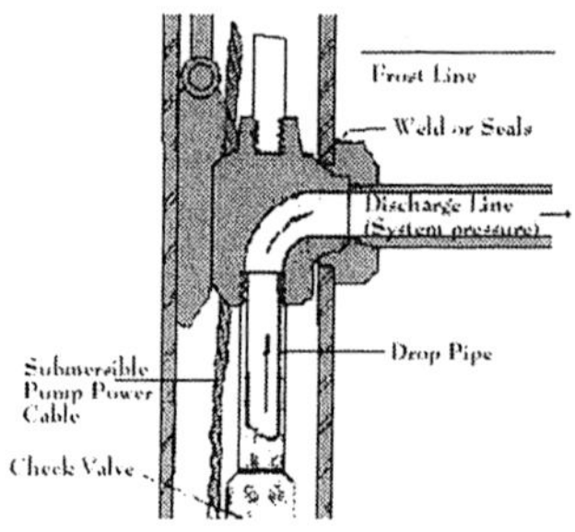

- **pitot tube**
  small ell shaped tube which can be attached to a pres-

sure gauge or other measuring device to measure the velocity head of water discharging from a nozzle or flowing in a pipe.

- **place of use**
  the specific location where water is applied or used. A water user cannot use water at another location without transferring the right or obtaining a new right.

- **placer deposit**
  a deposit of a mineral formed by a concentration of heavy minerals in flowing water, such as by a stream or waves.

- **plant available water**
  available water located in the root zone. Same as root zone available water.

- **plant classification**
  see **classification.**

- **plant water requirement**
  see **water requirement**.
  same as evapotranspiration.

- **playas**
  the flat-floored bottom of an undrained desert basin which may become a shallow lake.

- **plug**
  cement, grout or other material used to fill and seal a hole drilled for a water well.

- **plume**
  the area taken up by contaminant(s) in an aquifer.

- **pluvial**
  pertaining to precipitation.

- **point of connection**
  location where irrigation system is connected to a (potable) water supply.

- **point source**
  source of pollution that involves discharge of wastes from an identifiable point, such as a smokestack or sewage treatment plant.

- **polluter pays principle or residuals tax**
  a means of shifting the cost of pollution from the community to the polluter, usually in the form of a tax.

- **pollution**
  undesirable change in the physical, chemical or biological characteristics of the air, water or land that can harmfully affect the health, survival or activities of human or other living organisms.

- **pollution prevention**
  the elimination of potential pollutants at their source such as in a manufacturing process,

rather than at the point of discharge to the environment (end-of-pipe) or later.

- **pollution standards index (PSI)**
an index of air quality that is a combined measure of the health effects of several pollutants. It ranges from 0 (healthy) to 500 (extremely unhealthy).

- **poly vinyl chloride**
the real name for PVC. (See **PVC**)

- **Polychlorinated Biphenyls (PCBS)**
a group of synthetic, toxic industrial chemical compounds once used in making paint and electrical transformers, which are chemically inert and not biodegradable. PCBs were frequently found in industrial wastes and subsequently found their way into surface and ground waters. As a result of their persistence, they tend to accumulate in the environment. In terms of streams and rivers, PCBs are drawn to sediment, to which they attach and can remain virtually indefinitely. Although virtually banned in 1979 with the passage of the Toxic Substances Control Act, they continue to appear in the flesh of fish and other animals.

- **polyethylene**
a plastic used for manufacturing irrigation tubing. 'Poly' for short. Poly pipe is almost always black in colour, sometimes with a strip of a different colour for identification. It is very flexible and is usually sold in coils of 100 feet or more of tube. Poly pipe is often used in areas where the ground freezes 12" deep or more and also in mountainous areas that are extremely rocky. Poly pipe uses insert type fittings where a barbed shank is shoved into the end of the tubing. These fittings must be clamped, the barbs alone will not hole the tube on the fitting. Exception-special barbed fittings made for sprinkler risers do not need to be clamped. Also see **PVC for the most commonly used pipe material**.

- **polymer crystals**
inorganic soil additives that retain a significant quantity of water, thereby making mois-

ture available for future use by the surrounding vegetation.

- **pond**
  a body of water usually smaller than a lake and larger than a pool either naturally or artificially confined.

- **pondage.**
  small-scale storage at a waterpower plant to equalise daily or weekly fluctuations in river flow or to permit irregular hourly use of the water for power generation to accord with fluctuations in load.

- **pool**
  a deep reach of a stream. The reach of a stream between two riffles. Natural streams often consist of a succession of pools and riffles.

- **pop-up sprinkler head**
  a sprinkler head that retracts below ground level when it is not operating. Pop-up sprinklers that stick in the raised position are known as 'lawn mower food'.

- **porosity**
  a measure of the water-bearing capacity of subsurface rock. With respect to water movement, it is not just the total magnitude of porosity that is important but the size of the voids and the extent to which they are interconnected, as the pores in a formation may be open or interconnected or closed and isolated. For example, clay may have a very high porosity with respect to potential water content but it constitutes a poor medium as an aquifer because the pores are usually so small.

- **porous**
  something which allows water to pass through it.

- **portable (hand move) irrigation**
  see **irrigation system.**

- **positive displacement pump**
  any mechanism that seals water in a chamber, then forces it out by reducing the volume of the chamber. Examples: piston (including jack), diaphragm, rotary vane. Used for low volume and high lift. Synonyms: volumetric pump, force pump.

- **positive displacement pump**
  pump that moves a fixed quantity of fluid with each stroke or rotation, such as a piston or gear pump.

- **potable**
  suitable, safe or prepared for drinking.

- **potable water**
  treated water, which is meant for human consumption. It can be used for irrigation systems as long as protection is provided to prevent contamination of the domestic supply.

- **potential**
  amount of work that must be done per unit quantity of pure water in order to transport reversibly and isothermally an infinitesimal quantity of water from a pool of pure water at a specified elevation at atmospheric pressure to the soil water at the point under consideration. (ft, bar).

- **potential evapotranspiration**
  water loss that will occur if at no time there is a deficiency of water in the soil for use of vegetation.

- **potential natural water loss**
  the water loss during years when the annual precipitation greatly exceeds the average water loss. It represents the approximate upper limit to water loss under the type and density of vegetation native to a basin, actual conditions of moisture supply and other basin characteristics, whereas potential evapotranspiration represents the hypothetical condition of no deficiency of water in the soil at any time for use of the type and density of vegetation that would develop.

- **potential rate of evaporation**
  see **evaporativity**.

- **potential resource**
  a portion of the natural or human environment that is not today considered of value but that one day may gain value as a result of technological, cognitive or economic developments.

- **potentiometric surface**
  an imaginary surface representing the total head of ground water.

- **power**
  the rate at which work is done. It is the product of voltage times current, measured in Watts. 1000 Watts = 1 Kilowatt. An electric motor requires approximately 1 Kilowatt per Horsepower (after typical efficiency losses). 1 Kilowatt for 1 hour = 1 Kilowatt-hour (kWh).

- **Parts Per Billion**
  number of parts of a chemical found in one billion parts of a solid, liquid or gaseous mixture. Equivalent to micrograms per litre (μg/l).

- **Parts Per Million**
  number of parts of a chemical found in one million parts of a solid, liquid or gaseous mixture. Equivalent to milligrams per litre (mg/l).

- **precipitate**
  a solid which has come out of an aqueous solution. (e.g. iron from groundwater precipitates to a rust coloured solid when exposed to air).

- **precipitation**
  as used in hydrology, precipitation is the discharge of water, in liquid or solid state, out of the atmosphere, generally upon a land or water surface. It is the common process by which atmospheric water becomes surface or subsurface water .The term 'precipitation' is also commonly used to designate the quantity of water that is precipitated. Precipitation includes rainfall, snow, hail and sleet and is therefore a more general term than rainfall.

- **precipitation rate**
  a measurement of water application. The measurement is given in the depth of water applied to the soil. In other words the depth that the water would be if it didn't run-off or soak into the soil.

- **preservative**
  a chemical added to a water sample to keep it stable and prevent compounds in it from changing to other forms or to prevent micro-organism densities from changing prior to analysis.

- **pressure**
  the amount of force applied by water that is either forced by a pump or by the gravity.

Measured in pounds per square inch (psi). psi = vertical lift (or drop) in feet / 2.31 or .43 psi per foot.

- **pressure due to surge**
water pressure caused due to changes in water velocity in a pipe system. Also referred to as surge pressure.

- **pressure gauge**
a device used to measure water pressure.

- **pressure head**
measurement of water pressure based on the water depth. Measurement is stated as 'feet of head' or 'meters of head'. One foot of head is the pressure at the bottom of a 1 foot high column of water, which is also equal to 0.433 psi. So it's really a measure of the weight of water of a given depth. It doesn't matter how much water is present, the depth of the water only determines the pressure head. The water pressure at the bottom of a 2" diameter, 20 foot tall water filled pipe is the same as the water pressure at the bottom of a 20 foot deep lake.

- **pressure loss**
the term given for the loss of energy , in the form of pressure, that occurs whenever water moves through a pipe or any other piece of irrigation equipment. Pressure loss also occurs when water moves uphill against the force of gravity. If the total pressure loss in a piping system exceeds the available static water pressure the water will not flow. In landscape irrigation systems no flow means no grow.

- **pressure rating**
estimated maximum internal pressure that can be continuously exerted in a pipe or container with a high degree of certainty that it will not fail.

- **pressure regulator**
a device that regulates and ensures that the water pressure is kept within design limits for the drip irrigation system.

- **pressure switch**
an electrical switch actuated by the pressure in a pressure tank. When the pressure drops to a low set-point (cut-in) it turns a pump on. At a

high point (cut-out) it turns the pump off.

- **pressure tank**
  a fully enclosed tank with an air space inside. As water is forced in, the air compresses. The stored water may be released after the pump has stopped. Most pressure tanks contain a rubber bladder to capture the air. Synonym: captive air tank. Sizes range from less than 1 gallon up to several hundred. 25 to 120 gallons are the most common sizes.

- **pressure tank precharge**
  the pressure of compressed air stored in a captive air pressure tank. A reading should be taken with an air pressure gauge (tyre gauge) with water pressure at zero. The air pressure is then adjusted to about 3 psi lower than the cut-in pressure. If precharge is not set properly, the tank will not work to full capacity and the pump will cycle on and off more frequently.

- **pressure vacuum breaker (PVB)**
  1. backflow device configured with a spring loaded float and an independent spring loaded check valve.
  2. Check valve, which is designed to close with the aid of a spring when flow stops. It also has an air inlet valve, which is designed to open when the internal pressure is one psi above atmospheric pressure so that no non-potable liquid may be siphoned back into the potable water system. Being spring loaded it does not rely upon gravity as does the atmospheric vacuum breaker. This assembly includes resilient seated shut-off valves and test cocks. The PVB must be installed at least twelve (12) inches above all downstream piping and outlets. The PVB may be used to protect against a pollutant or contaminant. However, it may only be used to protect against backsiphonage. It is not acceptable protection against backpressure.

- **pressure**
  the force exerted over a surface divided by its area. In irrigation, pressure usually describes the amount of en-

ergy available to move water through a pipe, sprinkler or emitter.

- **primary standards**
air pollution standards designed to protect human health.

- **primary treatment**
sewage treatment consisting of removal of solids by sedimentation, flocculation, screening or similar methods.

- **primary wastewater treatment**
the first stage of the wastewater-treatment process where mechanical methods, such as filters and scrapers, are used to remove pollutants. Solid material in sewage also settles out in this process.

- **priming**
the process of hand-filling the suction pipe and intake of a surface pump. Priming is generally necessary when a pump must be located above the water source. A 'self-priming' pump is able to draw some air suction in order to prime itself, at least in theory. See **foot valve**.

- **principle of limiting factors**
whatever factor (nutrient, water, sunlight, etc.) is in shortest supply will limit the growth and development of an organism or a community.

- **procedural equity**
a situation in which there is a differential application of environmental regulations, laws or treaties such that some areas bear more pollution burdens than others.

- **process water**
water used for nonpotable industrial usage, e.g., mixing cement.

- **profile (soil)**
vertical section of the soil through all its horizons and extending into the C horizon.

- **profile (sprinkler)**
chart showing the application rates vs. distance of throw for a sprinkler head. See also **single leg profile**.

- **profundal zone**
a lake's deep-water region that is not penetrated by sunlight.

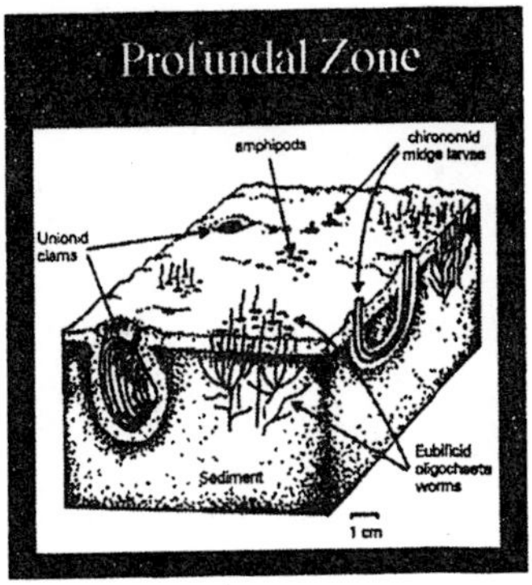

- **propeller pump**
pump which develops most of its head by the lifting action of vanes on the water.

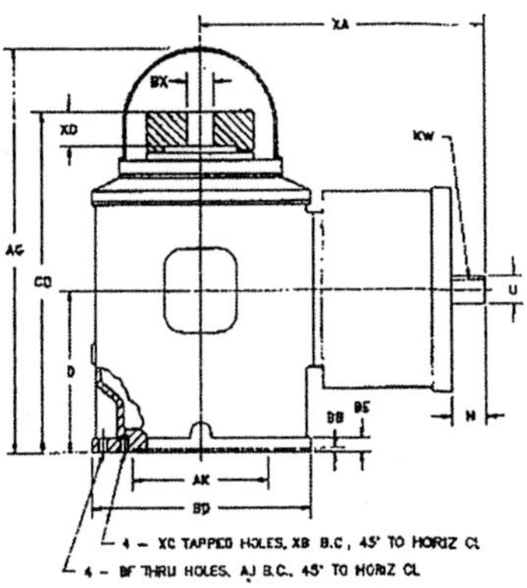

- **proxy value**
a price applied to a commodity that has no established market value.

- **psi**
abbreviation used by irrigation designers when they don't have enough room to write 'pounds per square inch'.

- **public supply**
water withdrawn by public governments and agencies, such as a country water department and by private companies that is then delivered to users. Public suppliers provide water for domestic, commercial, thermoelectric power, industrial and public water users. A public water supplier delivers most people's household water.

- **Public Water Supply (PWS) utilities**
utilities that provide potable water for public use.

- **public water use**
water supplied from a public-water supply and used for such purposes as firefighting, street washing and municipal parks and swimming pools.

- **puddle**
a small pool of water, usually a few inches in depth and from several inches to several feet in its greatest dimension.

- **pulsation damper**
  a device that absorbs and releases pulsations in flow produced by a piston or diaphragm pump. Consists of a chamber with air trapped within it. A 'shock absorber' to prevent water hammers and to provide a more even pressure.

- **pump**
  a device which increases the water pressure or moves water. Technically most pumps don't move the water, they increase the water pressure and the water pressure moves the water.

- **pump column**
  pipe through which water from well pumps (vertical turbine impellers) is conveyed to the ground surface (pump discharge head).

- **pump controller**
  an electronic device which varies the voltage and current of a PV array to match the needs of an array-directed pump. It allows the pump to start and to run under low sun conditions without stalling. Electrical analogy: variable transformer. Mechanical analogy: automatic transmission.

- **pump jack**
  a deep well piston pump. The piston and cylinder is submerged in the well water and actuated by a rod inside the drop pipe, powered by a motor at the surface. This is an old-fashioned system still used for extremely deep wells, including solar pumps as deep as 1000 feet. In solar powered systems, a DC motor replaces the windmill.

- **pump start circuit**
  feature on automatic controllers which supplies 24 VAC, which can be used to activate a pump through an external pump start relay.

- **pump start relay**
  low-amperage or electric switch designed for use with pump start circuits.

- **pump-back system**
  a return flow system in which tailwater is pumped back to the head of an irrigation ditch for reuse.

- **pumped hydroelectric storage**
  storing water for future use in generating electricity. Excess electrical energy pro-

duced during a period of low demand is used to pump water up to a reservoir. When demand is high, the water is released to operate a hydroelectric generator.

- **pumping lift**
  distance water must be lifted in a well from the pumping level to the ground surface.

- **pumping plant**
  facility that lifts water up and over hills.

- **purge**
  to force a gas through a water sample to liberate volatile chemicals or other gases from the water so their level can be measured.

- **purgeable organics**
  volatile organic chemicals which can be forced out of the water sample with relative ease through purging.

- **PV**
  the common abbreviation for photovoltaic.

- **PV array**
  a group of PV (photovoltaic) modules (also called panels) arranged to produce the voltage and power desired.

- **PV array-direct**
  the use of electric power directly from a photovoltaic array, without storage batteries to store or stabilise it. Most solar water pumps work this way, utilising a tank to store water.

- **PV cell**
  the individual photovoltaic device. The most common PV modules are made with 33 to 36 silicon cells each producing about 1/2 volt.

- **PV module**
  an assembly of PV cells framed into a weatherproof unit. Commonly called a 'PV panel' or solar panel. See **PV array**. Most solar crystalline solar panels have 36 cells, most thin film panels have 33.

- **PVC**
  abbreviation for poly-vinyl-chloride. A type of plastic used to make water pipe. Usually white in colour but sometimes is grey, brown, tan or purple.

- **quad**
  a measure of energy use, equal to one quadrillion (1,000,000,000,000,000) British thermal units.

- **quarry water**
  the moisture content of freshly quarried stone, especially if porous.

- **quicksilver water**
  a solution of mercury nitrate used in gilding.

- **quickwater**
  the part of a stream that has a strong current. An artificial current or bubbling patch of water just astern of a moving boat.

- **RPA**
  abbreviation for reduced pressure assembly. See **reduced pressure**.

- **radial flow pump**
  see **pump**.

- **radiation inversion**
  a temperature inversion caused by radiational cooling of air close to the ground.

- **radius of influence**
  the radial distance from the centre of a well bore to the point where there is no lowering of the water table or potentiometric surface.

- **radius of throw**
  distance from the sprinkler head to the furthest point of water application and is used only for heads with circular wetting patterns.

- **rain**
  water drops which fall to the earth from the air.

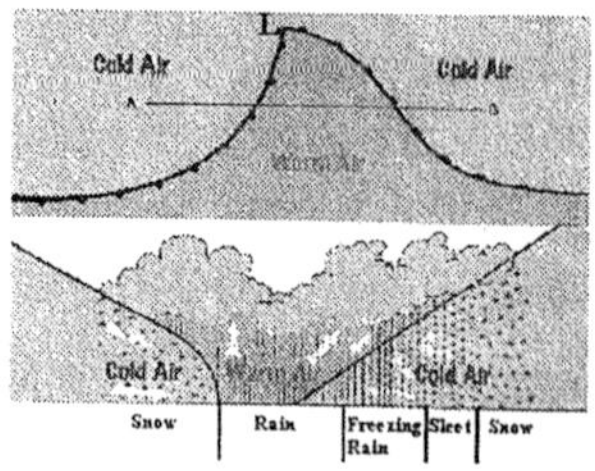

- **rain gauge**
  any instrument used for recording and measuring time, distribution and the amount of rainfall.

- **rain sensor**
  device to measure rainfall.

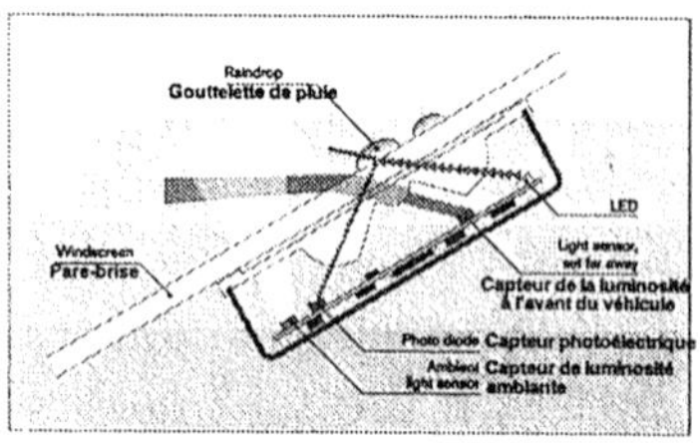

- **rain shut-off device**
  device which prevents voltage from the controller from activating the valves when a preset amount of rain falls.

- **rain switch**
  see **rain shut-off device**.

- **rainfall**
  the quantity of water that falls as rain only.

- **rainfall excess**
  the volume of rainfall available for direct runoff. It is equal to the total rainfall minus interception, depression storage and absorption.

- **rainfall intensity**
  rate of rainfall for the design storm frequency and for the time of concentration of the drainage area. (Used in rational method to compute runoff.)

- **rainfall, excessive**
  rainfall in which the rate of fall is greater than certain adopted limits, chosen with regard to the normal precipitation (excluding snow) of a given place or area

- **range condition**
  an estimate of the degree to which the present vegetation and ground cover depart from that which is presumed to be the natural potential (or climax) for the site.

- **rangeland**
  land that provides or is capable of providing forage for grazing animals.

- **rapid-rate exfiltration basin (REB)**
  an artificial impoundment that provides for fluid losses through percolation/seepage as well as through evaporative losses.

- **rating curve**
  a drawn curve showing the relation between gauge height and discharge of a stream at a given gauging station.

- **rational equation**
  equation used to predict the runoff from a watershed.

- **reach**
  1. The length of channel uniform with respect to discharge, depth, area and slope.
  2. The length of a channel for which a single gauge affords a satisfactory measure of the stage and discharge.
  3. The length of a river between two gauging stations.
  4. More generally, any length of a river.

- **readily available water**
  1. is the actual amount of water in the soil that is allowed to be removed from the soil prior to irrigation.

Readily available water (RAW) is a function of MAD and the available water (amount between field capacity and wilting point). The available water (AW) is a function of the effective root depth and the available water per foot in the soil. (RAW = MAD * AW)
2. portion of available water that is more readily available for plant usage. It varies with plant type.

- **receiving waters**
a river, ocean, stream or other watercourse into which wastewater or treated effluent is discharged.

- **recession curve**
a hydrograph showing the decreasing rate of runoff following a period of rain or snowmelt. Since direct runoff and base runoff recede at different rates, separate curves, called direct runoff recession curves or base runoff recession curves, are generally drawn. The term depletion curve' in the sense of base runoff recession is not recommended.

- **recharge**
Increases in groundwater storage from precipitation, infiltration from streams or human activity (artificial recharge), such as putting surface water into spreading basins moisture the air could hold under the same conditions, usually expressed as a percentage.

- **recharge area**
area in which water reaches the zone of saturation by surface infiltration. An area in which there are downward components of hydraulic head in the aquifer. Infiltration moves downward into the deeper parts of an aquifer in a recharge area.

- **recharge zone**
the area where a formation allows available water to enter the aquifer.

- **reclaimed wastewater**
treated wastewater that can be used for beneficial purposes, such as irrigating certain plants.

- **reclaimed water**
domestic wastewater that is under the direct control of a treatment plant owner/

operator which has been treated to a quality suitable for a beneficial use.

- **reclamation**
  the process of matter or energy being reclaimed or restored to a better or useful state. For instance, dams reclaim water, which would otherwise be lost through runoff.

- **rectangular spacing**
  see **spacing.**

- **recurrence interval**
  average amount of time between events of a given magnitude. For example, there is a 1% chance that a 100-year flood will occur in any given year.

- **recurrence interval (return period).**
  the average interval of time within which the given flood will be equalled or exceeded once.

- **recycled water**
  water that is used more than one time before it passes back into the natural hydrologic system.

- **recycling**
  reprocessing of a used product for reuse in a similar or different form.

- **reduced pressure principle assembly (RP, RPA, RPZ)**
  consists of two internally loaded independently operating check valves and a mechanically independent, hydraulically dependent relief valve located between the check valves. This relief valve is designed to maintain a zone of reduced pressure between the two check valves at all times. The RP also contains tightly closing, resilient seated shut-off valves upstream and downstream of the check valves along with resilient seated test cocks. This assembly is used for the protection of the potable water supply from either pollutants or contaminants and may be used to protect against either backsiphonage or backpressure.

- **reducer**
  a fitting used to change from one size pipe to another. Two types are generally available. The first, and most common is the reducer bushing (abbreviated 'red bush' or RB). The reducer bushing fits inside a coupling or other fitting on the large end. The pipe fits

into the red bush on the small end.. The other common reducer is a 'bell reducer'.

- **reference crop evapotranspiration**
  see **evapotranspiration**.

- **reflected light**
  light that reaches a plant after bouncing off surrounding surfaces. See **micro-climate.**

- **regime**
  'regime theory' is a theory of the forming of channels in material carried by the streams. As used in this sense, the word 'regime' applies only to streams that make at least part of their boundaries from their transported load and part of their transported load from their boundaries, carrying out the process at different places and times in any one stream in a balanced or alternating manner that prevents unlimited growth or removal of boundaries. A stream, river or canal of this type is called a 'regime stream, river or canal'. A regime channel is said to be 'in regime' when it has achieved average equilibrium, that is, the average values of the quantities that constitute regime do not show a definite trend over a considerable period—generally of the order of a decade. In unspeci-alised use 'regime' and 'regimen' are synonyms.

- **regimen of a stream**
  the system or order characteristic of a stream. In other words, its habits with respect to velocity and volume, form of and changes in channel, capacity to transport sediment and amount of material supplied for transportation. The term is also applied to a stream which has reached an equilibrium between corrosion and deposition or in other words, to a graded stream.

- **regulation**
  the artificial manipulation of the flow of a stream.

- **relative humidity**
  ratio of the amount of water vapour present in the atmosphere to the amount required for saturation at the same dry bulb temperature.

- **relative scarcity**
  short supply of a resource in one or more areas due to inadequate or disrupted distribution.

- **relay**
  electrical or electronic device which uses a signal current to actuate a separate electrical circuit.

- **renewable energy**
  energy resources that are produced naturally as fast as they are consumed, such as solar, wind and hydroelectric power.

- **renewable resource**
  a resource that can be depleted but will be replenished by natural processes. Forests and fisheries are examples.

- **replacement cost**
  the cost of replacing a resource that is used.

- **re-regulating reservoirs**
  a reservoir for reducing diurnal fluctuations resulting from the operation of an upstream reservoir for power production.

- **reserve**
  1. in the context of mineral resources, a deposit of known location and quality that is economically extractable at the present time.
  2. amount of a particular resource in known locations that can be extracted at a profit with present technology and prices.

- **reservoir**
  a pond, lake, tank or basin (natural or human made) where water is collected and used for storage. Large bodies of groundwater are called groundwater reservoirs, water behind a dam is also called a reservoir of water.

- **residual**
  waste products of production. Air pollution is an example of residuals from manufacturing.

- **residual chlorine**
  the available chlorine which remains in solution after the demand has been satisfied.

- **residual tax**
  a tax on production based on the amount of waste produced.

- **residual-mass curve**
  a graph of the cumulative departures from a given reference such as the arith-

metic average, generally as ordinate, plotted against time or date, as abscissa.

- **residuals management**
  an approach to production of goods and services that includes accounting of waste products associated with both production and consumption phases.

- **resource**
  something that is useful to humans.

- **resource recovery**
  separation of waste into recyclable components such as metal, glass and heat from incineration.

- **retarding reservoir**
  ungated reservoir for temporary storage of flood water. Sometimes called detention reservoir.

- **retraction**
  operation when the pop-up riser of a sprinkler such as a spray head or rotor returns to the case in the ground. Also called pop-down.

- **retrofitting**
  the replacement of existing water fixtures, appliances and devices with more efficient fixtures, appliances and devices for the purpose of water conservation.

- **return flow**
  1. that part of a diverted flow that is not consumptively used and is returned to its original source or another body of water.
  2. drainage water from irrigated farmlands that re-enters the water system to be used further downstream.

- **return-flow system**
  a system of pipelines or ditches to collect and convey surface or subsurface runoff from an irrigated field for reuse. Sometimes called a 'reuse system.'

- **reuse**
  repeated use of a product without reprocessing or remanufacture.

- **reverse osmosis**
  a water treatment method whereby water is forced through a semipermeable membrane which filters out impurities.

- **Reynolds number**
  dimensionless number used to characterise flow of water.

- **rhizome**
a thickened, modified stem that grows horizontally along or under the soil surface. An example is an iris.

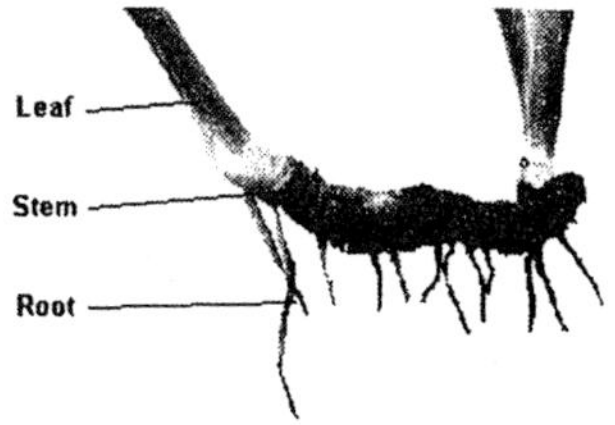

- **riffle**
a rapid in a stream.

- **right of free capture**
the idea that the water under a person's land belongs to that person and they are free to capture and use as much as they want.

- **rill**
a small channel created by soil erosion and small enough to be obliterated by ploughing.

- **riparian areas**
lands adjacent to and subject to flooding by streams.

- **riparian zone**
a stream and all the vegetation on its banks.

- **riser**
1. the connection between a sprinkler or other irrigation device and the pipe that supplies the water to it, usually consisting of a short nipple and sometimes a few ells.
2. length of pipe which has male nominal pipe threads on each end and is usually affixed to a lateral or submain to support a sprinkler or anti-siphon valve.

- **river**
a natural stream of water of considerable volume, larger than a brook or creek.

- **river authority**
any district or authority created by the legislature and which is governed by a board of directors appointed or designated in whole or in part by the governor

- **river basin**

  the area drained by a river and its tributaries.

- **rock garden**

  in landscape terms, a garden with rock outcroppings, low-growing groundcovers and perennials. Rock garden plants generally prefer the

hot, dry conditions produced by the surrounding rocks.

- **root depth**

  depth from which roots extract water. The effective rooting depth is generally the depth from which the crop is currently capable of extracting soil water. However, it may also be expressed as the depth from which the crop can extract water when mature or the depth from which a future crop can extract soil water. Maximum effective root depth depends on the rooting capability of the plant, soil profile characteristics and moisture levels in the soil profile.

- **root zone**

  depth of soil that plants roots readily penetrate and in which the predominant root activity occurs. Area of the soil from which the crop roots extract water and nutrients. Crop rooting depth is typically taken as the soil depth containing 80 percent of plant roots. (May also be used as a portion of the root zone in equations where soil characteristics change within the root zone.)

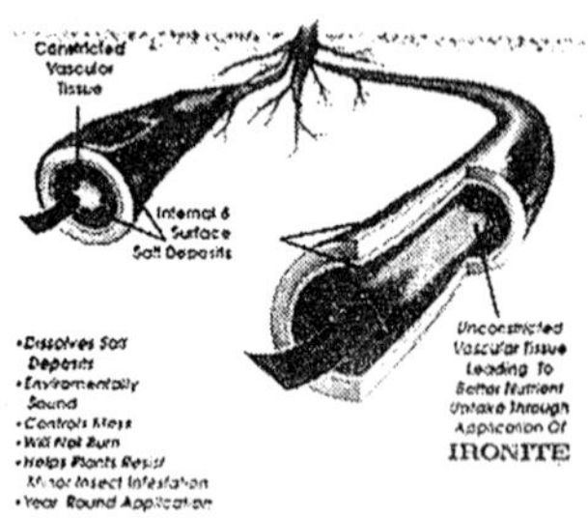

- **rotational speed**

  number of revolutions made per unit time (usually minutes) by a shaft or other object.

- **ruminant**
  one of a group of grazing animals including cattle, bison, sheep, goat, which have digestive systems particularly adapted to grasses.

- **run time**
  length of time to operate an irrigation system or an individual zone

- **run time multiplier**
  multiplier used to compensate for the lack of perfect uniformity in a sprinkler system. Also, see **scheduling coefficient**.

- **runner**
  a slender stem sent out from the base of certain perennials from which a new plant forms. An example is strawberry.

- **runoff**
  that part of the precipitation that appears in surface streams. It is the same as streamflow unaffected by artificial diversions, storage or other works of man in or on the stream channels. Runoff may be classified as follows:
  Classification as to speed of appearance after rainfall or snow melting:
  Direct runoff
  Base runoff
  Classification as to source:
  Surface runoff (see **Overland flow)**
  Storm seepage
  Ground-water runoff

- **runoff rate**
  rate at which water flows above ground from a watershed or field. Water, which is not absorbed by the soil and drains to another location. Runoff occurs when water is applied in excessive amounts or too quickly.

- **runout**
  see **water yield**

- **safety rope**
  plastic rope used to suspend the pump, primarily in case of pipe breakage.

- **sahel**
  a semiarid east-west swath across Africa, environmentally transitional between the Sahara Desert (to the north) and equatorial rainforests (to the south), in which recent desertification and drought have been particularly severe.

- **sail index**
  cross-sectional area per unit volume of water of the drop comprising the sprinkler spray. Cross sectional area of spray (drops) per unit volume of water.

- **saline**
  the condition of containing dissolved or soluble salts. Saline soils are those whose productivity is impaired by high soluble salt content. Saline water is that which would impair production if used to irrigate sensitive crops without adequate leaching to prevent soil salinisation.

- **saline soil**
  non-sodic soil containing soluble salts in such quantities that they interfere with the growth of most crops .Soil that has sufficient soluble salts to interfere with crop growth. Non-sodic soil containing sufficient soluble salts to impair its productivity for growing most crops. The electrical conductivity (ECe) of the saturation extract is greater than 4 mmhos/cm and exchangeable sodium percentage (ESP) is less than 15 i.e., non-sodic. The principal ions are chloride, sulphate, small amounts of bicarbonate and occasionally some nitrate. Sensitive plants are affected at half this salinity and highly tolerant ones at about twice this salinity.

- **saline water**
  water that contains significant amounts of dissolved solids. Here are our parameters for saline water:
  Fresh water - Less than 1,000 parts per million (ppm). Slightly saline water - From 1,000 ppm to 3,000 ppm. Moderately saline water - From 3,000 ppm to 10,000 ppm. Highly saline water - From 10,000 ppm to 35,000 ppm.

- **saline-sodic soil**
  soil containing both sufficient soluble salts and exchangeable sodium to interfere with the growth of most crops. The exchangeable sodium percentage (ESP) is greater than or equal to 15 and electrical conductivity of the saturation extract (ECE) is greater than 4 mmhos/cm. It is difficult to

leach because the clay colloids are dispersed.

- **salinity**
the concentration of mineral salts in water. The average salinity of the oceans is about 35 parts per thousand.

- **saltwater intrusion**
movement of salt water into aquifers formerly occupied by fresh water as a result of groundwater withdrawal in coastal areas.

- **sand**
soil particles ranging from 50 to 200 μ m (0.05 to 0.2 mm) in diameter.
Soil material containing 85% or more in this size range.

- **sandy soil**
determined by particle size. Sandy soils have large particles, are loosely arranged and do not hold water.

- **sanitary landfill**
landfill that is lined with plastic or concrete or located in clay-rich soils to prevent hazardous substances from leaking into the environment.

- **SAS**
surficial aquifer system.

- **satisficing**
a decision-making strategy that seeks a course of action that is good enough but not necessarily perfect. A few alternatives are compared and the best course of action is chosen from this limited range of options.

- **saturation**
the condition of a liquid when it has taken into solution the maximum possible quantity of a given substance at a given temperature and pressure.

- **savanna**
tropical or subtropical semiarid grassland with scattered trees.

- **SCH 40, SCH 80**
means 'schedule 40, 80'. Sometimes abbreviated 'SCHED'. A set of standards for pipe diameter and wall thickness used for both plastic and steel pipe. SCH 40 pipe is the standard type used for water pipes sized 1 1/2 and smaller. SCH 80 is a similar standard having thicker walls than SCH 40.

- **schedule**
method of specifying the dimensions and thus the allow-

able operating pressure of pipe. For a specific schedule rating, the wall thickness remains relatively constant for different pipe diameters, resulting in lower allowable operating pressures for larger diameter pipe.

- **schedule, irrigation**
  see **irrigation schedule**.

- **scheduling**
  determining when to irrigate and how much water to apply, based upon measurements or estimates of soil moisture or crop water used by a plant.

- **scheduling coefficient**
  1. a number that relates to the uniformity of coverage and how to operate the system to adequately irrigate the entire turf area. It indicates the amount of extra watering time needed to adequately irrigate the driest areas.
  2. Numerical expression, which serves as an index of the uniformity of water application to a give area within a specific geometric arrangement of sprinklers (e.g. triangular or square). Used to measure the uniformity of landscape irrigation systems.

- **scheduling programme**
  set of specifications identifying times to turn on and off water to various zones of an irrigation system.

- **SDR**
  see **standard dimension ratio**.

- **seal**
  the impermeable material, such as cement grout bentonite or puddling clay placed in the annular space between the borehole wall and the casing of a water well to prevent the downhole movement of surface water or the vertical mixing of artestian waters.

- **sealed piston pump**
  a type of pump in which water is drawn in and forced out of a chamber by a piston mechanism. The pistons have a very short stroke, allowing the use of flexible gaskets to seal water out of the piston mechanism. Check valves let water into and out of the chamber.

- **secondary standard**
  an air-quality standard designed to protect human wel-

fare (property, vegetation, etc.) as opposed to health.

- **secondary treatment**
second step in most waste treatment systems, in which bacteria break down the organic parts of sewage wastes, usually accomplished by bringing the sewage and bacteria together in trickling filters or in the activated sludge process. Compare **primary treatment, tertiary treatment. Compare primary treatment, tertiary treatment.**

- **secondary wastewater treatment**
treatment (following primary wastewater treatment) involving the biological process of reducing suspended, colloidal and dissolved organic matter in effluent from primary treatment systems and which generally removes 80 to 95 percent of the Biochemical Oxygen Demand (BOD) and suspended matter. Secondary wastewater treatment may be accomplished by biological or chemical-physical methods. Activated sludge and trickling filters are two of the most common means of secondary treatment. It is accomplished by bringing together waste, bacteria and oxygen in trickling filters or in the activated sludge process. This treatment removes floating and settleable solids and about 90 percent of the oxygen-demanding substances and suspended solids. Disinfection is the final stage of secondary treatment.

- **second-foot**
same as CFS.

- **sedentarisation**
permanent settlement of once-nomadic people.

- **sediment**
soil particles, sand and minerals washed from the land into aquatic systems as a result of natural and human activities.

SEDIMENT ISOPACH MAP OF THE NEPONSET RIVER
ABOVE THE TILESTONE & HOLLINGSWORTH DAM
Milton / Hyde Park, Massachusetts

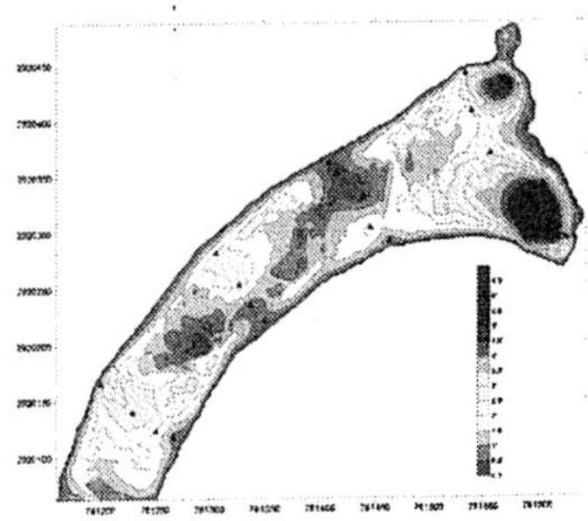

- **sediment discharge**
  the rate at which dry weight of sediment passes a section of a stream or is the quantity of sediment, as measured by dry weight or by volume, that is discharged in a given time.

- **sediment load**
  amount of sediment carried by running water.

- **sedimentary cycle**
  biogeochemical cycle in which materials primarily are moved from land to sea and back again.

The Sedimentary Cycle

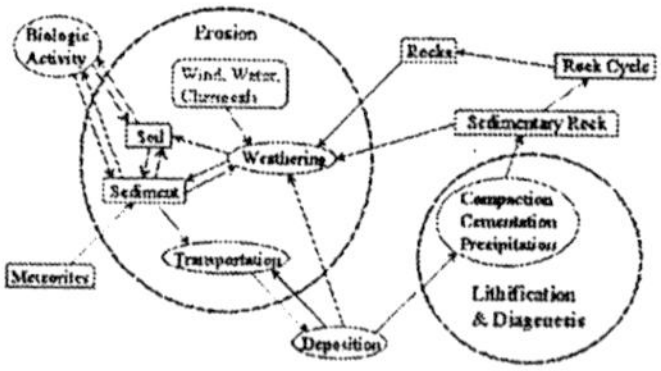

- **sedimentary rock**
  rock formed of sediment and specifically:
  1. sandstone and shale, formed of fragments of other rock transported from their sources and deposited in water and
  2. rocks formed by or from secretions of organisms, such as most limestone.

  Many sedimentary rocks show distinct layering, which is the result of different types of sediment being deposited in succession.

- **sedimentation**
  deposition of waterborne sediments due to a decrease in velocity and corresponding reduction in the size and amount of sediment which can be carried.

- **sedimentation tanks**
  wastewater tanks in which floating wastes are skimmed off and settled solids are removed for disposal.

- **seep**
  a spot where water contained in the ground oozes slowly to the surface and often forms a pool. A small spring.

- **seepage**
  1. the slow movement of water through small cracks, pores, interstices, etc., of a material into or out of a body of surface or subsurface water.
  2. the loss of water by infiltration into the soil from a canal, ditches, laterals, wa-

tercourse, reservoir, storage facilities or other body of water or from a field.

- **seepage irrigation systems**
irrigation systems which convey water through open ditches. Water is either applied to the soil surface (possibly in furrows) and held for a period of time to allow infiltration or is applied to the soil subsurface by raising the water table to wet the root zone.

- **seiche**
the free oscillation of the bulk of water in a lake and the motion caused by it on the surface of the lake.

- **self-supplied water**
water withdrawn from a surface or ground-water source by a user rather than being obtained from a public supply. An example would be homeowners getting their water from their own well.

- **semi-arid climate**
a climate characterised by 10 inches to 20 inches of annual rainfall.

- **semi-closed irrigation systems**
irrigation systems which convey water through closed pipes and distribute it to the crop through open furrows between crop rows.

- **semi-confining layers**
layers with little or no vertical flow that can store ground water and also transmit it slowly from one aquifer to another. The rate of vertical flow is dependent on the head differential between the semi-confining beds and those above and below them.

- **separate impacts**
effects of a system's activity that can be measured separately.

- **separate sewer**
a sewer system that carries only sanitary sewage, not storm water runoff. When a sewer is constructed this way, wastewater treatment plants can be sized to treat sanitary wastes only and all of the water entering the plant receives complete treatment at all times. Compare **combined sewer**.

- **septic tank**
a tank used to detain domestic wastes to allow the settling of solids prior to distribution to a leach field for

soil absorption. Septic tanks are used when a sewer line is not available to carry them to a treatment plant. A settling tank in which settled sludge is in immediate contact with sewage flowing through the tank and wherein solids are decomposed by anaerobic bacterial action.

- **service line**
section of piping connecting larger municipal supply line with water meter.

- **set time**
elapsed time between the beginning and end of water application to an irrigation set. Amount of time required to apply a specific amount of water during one irrigation to a given area, typically refilling the plant root zone to field capacity minus expected rainfall.

- **settleable solids**
in sewage, suspended solids that will settle when the sewage is brought to a quiet state for a reasonable length of time, usually two hours.

- **settling pond (water quality)**
an open lagoon into which wastewater contaminated with solid pollutants is placed and allowed to stand. The solid pollutants suspended in the water sink to the bottom of the lagoon and the liquid is allowed to overflow out of the enclosure.

- **sewage treatment plant**
a facility designed to receive the wastewater from domestic sources and to remove materials that damage water quality and threaten public health and safety when discharged into receiving streams or bodies of water. The substances removed are classified into four basic areas:
1. greases and fats,
2. solids from human waste and other sources,
3. dissolved pollutants from human waste and decomposition products and
4. dangerous micro-organisms.
Most facilities employ a combination of mechanical removal steps and bacterial decomposition to achieve the desired results. Chlorine is often added to discharges from the plants to reduce

the danger of spreading disease by the release of pathogenic bacteria.

- **sewer**
  a system of underground pipes that collect and deliver wastewater to treatment facilities or streams.

- **shadow price**
  an artificial monetary value applied to those resources for which a simple price tag is not easy to calculate, for example, wilderness, habitat.

- **shale oil**
  see also **kerogen**. A petroleum-like substance found in high concentrations in some shale rocks.

- **sheet wash**
  flow of rainwater that covers the entire ground surface with a thin film and is not concentrated into streams.

- **shelter belt**
  trees and/or shrubs planted in rows or groups to provide shelter and act as a windbreak to protect crops or ornamental plants. A shelter belt needs to be dense at the bottom to produce a good effect, as wind is blocked three feet from the bottom of the plants for every foot in height of the shelter belt. See **micro-climate**.

- **shelterwood cutting**
  a two-phase timber-harvesting technique in which not all trees are taken in the first phase so that some trees may provide shelter for young seedlings, when these are established, the remaining older trees are cut.

- **shrub sprinkler head**
  a sprinkler head mounted above ground level on a pipe, usually used for watering shrubs.

- **shutoff head**
  pressure head on the outlet side of a pump at which the discharge drops to zero. Maximum pressure a pump will develop at a given speed.

- **silt**
  mineral portion of soil with particle sizes ranging from 0.05 to 0.002 mm in diameter.

- **siltation**
  the deposition of finely divided soil and rock particles upon the bottom of stream and river beds and reservoirs.

- **single leg profile**
  precipitation rate profile of an individual sprinkler head operating at a known, constant pressure.

- **sinkhole**
  a depression in the Earth's surface caused by dissolving of underlying limestone, salt or gypsum. Drainage is provided through underground channels that may be enlarged by the collapse of a cavern roof.

- **siphon**
  closed conduit used to convey water across localised minor elevation raises in grade. It generally has end sections below the middle section. A vacuum pump is commonly used to remove air and keep the siphon primed. The upstream end must be under the water surface. Both ends must be under water or the lower end must be closed to prime the siphon.

- **siphon tube**
  relatively short, lightweight, curved tube used to convey water over ditch banks to irrigate furrows or borders.

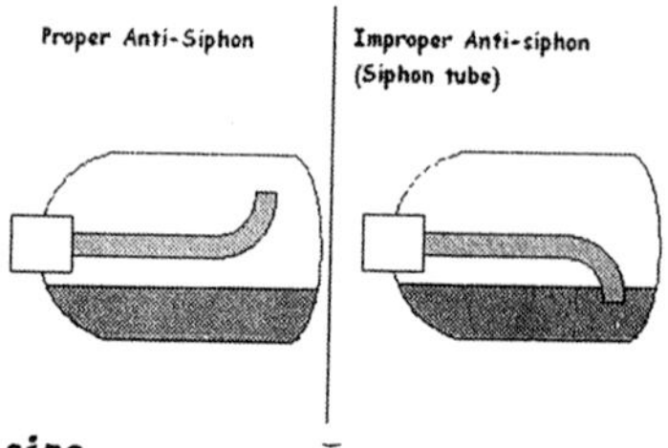

- **size**
  a higher number indicates smaller wire. Elsewhere it is specified in millimetres. The specification 'two-wire plus ground' will indicate three wires (conductors) in the cable. It is connected to a pump by splicing.

- **skimming**
  the diversion of water from a stream or conduit by a shallow overflow used to avoid diversion of sand, silt or other debris carried as bottom load.

- **sleet**
  precipitation which is a mixture of rain and ice.

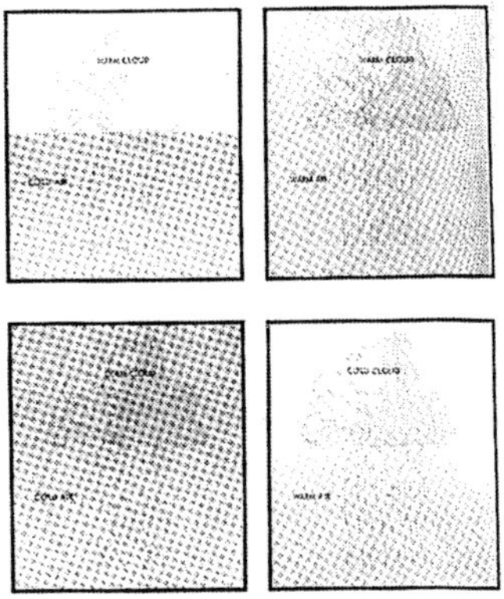

- **slickensides**
  a smooth striated polished surface produced on rock by movement along a fault.

- **slide gate**
  head control valve, which slides on rails, used to control drainage or irrigation water.

- **slip**
  slip is the term used to describe a solvent welded connection on a fitting. Example, 'hand me that slip-thread ell'. (That's an ell that has threads on one side and a glued socket on the other.)

- **slip configuration or slip joint**
  connection without threads (of PVC pipe or fittings) which is solvent welded.

- **slope**
  degree of deviation of a surface from the horizontal, usually expressed in percent or degrees.

- **slough**
  a channel in which water moves sluggishly or a place of deep muck, mud or mire. Sloughs are wetland habitats that serve as channels for water draining off surrounding uplands and/or wetlands.

- **sludge**
  solid matter that settles to the bottom of sedimentation tanks in a sewage treatment plant and must be disposed of by digestion or other methods or recycled to the land.

- **smog**
  a term used to describe air pollution.

- **snow**
  precipitation in the form of branched hexagonal crystals, often mixed with simple ice crystals, which fall more or less continuously from a solid cloud sheet. These crystals may fall either separately or in cohesive clusters forming snowflakes.

- **snow course**
  a line laid out and permanently marked on a drainage area along which the snow is sampled at definite distances or stations. Sampling occurs at appropriate times to determine snow depth, water equivalent and density.

- **snow density**
  ratio between the volume of melt water derived from a sample of snow and the initial volume of the sample. This is numerically equal to the specific gravity of the snow.

- **snow field**
  an area, usually at high elevation or in polar latitudes, where snow accumulates and remains throughout the entire year.

- **snow survey**
  the process or operation of determining the depth, water content and density of snow at various selected points on a drainage basin. This is done in order to determine the amount of water stored there in the form of snow for the purpose of forecasting runoff.

- **snow, quality of**
  the ratio of heat of melting of snow, in calories per gram to the 80 calories per gram for melting pure ice at 0°C. Percentage by weight which is ice.

- **snowfall**
  the amount of snow, hail, sleet or other precipitation occurring in solid form which reaches the earth's surface. It may be expressed in depth in inches after it falls or in terms of inches in depth of the equivalent amount of water.

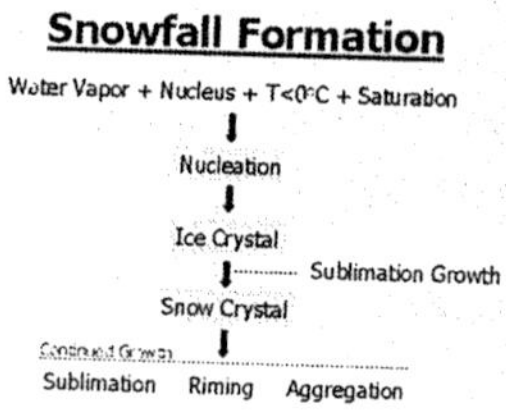

- **snowline**
  the general altitude to which the continuous snow cover of high mountains retreats in summer, chiefly controlled by the depth of the winter snowfall and by the temperature of the summer.

- **snowline, temporary**
  a line sometimes drawn on a weather map during the winter showing the southern limit of the snow cover.

- **snowpack**
  the winter accumulation of snow in mountain areas.

- **social cost**
  the cost of producing a good or service, plus its cost to humans in terms of pollution and other negative socio-environmental effects.

- **social equity**
  situations in which there are not different pollution burdens or access to resources based on race, ethnicity, age or gender.

- **socket**
  a socket is a female connection on a fitting. It can be threaded or glued, but most of the time the term is used for glued fittings.

- **sodic soil**
  non-saline soil containing sufficient exchangeable sodium to adversely affect crop production and soil structure.

- **sodium adsorption ratio (SAR)**
  portion of soluble sodium ions in relation to the soluble calcium and magnesium ions in the soil water extract .

- **sodium adsorption ratio, adjusted**
  sodium adsorption ratio of a water adjusted for the precipitation or dissolution of CA^{2+} and Mg^{2+} that is expected to occur where a water reacts with alkaline earth carbonates with a soil.

- **sodium percentage**
  percentage of total cations that is sodium in water or soil solution.

- **softwoods**
  timber species with relatively low-density wood, primarily needle-leaf trees.

- **soil**
  unconsolidated minerals and material on the immediate surface of the earth that serves as a natural medium for the growth of plants.

- **soil amendment**
  organic and inorganic materials added to soil to improve texture, nutrients, moisture-holding capacity and infiltration rates.

- **soil classification**
  systematic arrangement of soils into classes of one or more categories or levels to meet a specific objective. Broad groupings are made on the basis of general characteristics and subdivisions are made on the basis of more detailed differences in specific properties.

- **soil compaction**
  compression of soil that eliminates air pockets from the soil structure, resulting in the suffocation of the plant roots. Soil compaction may be caused by ongoing pedestrian traffic, one time or ongoing vehicular traffic, construction equipment or storage of materials.

- **soil conservation**
  protection of soil against physical loss by erosion and chemical deterioration by the application of management and land-use methods that safeguard the soil against all natural and human-induced factors.

- **soil erodibility**
  a measure of the inherent susceptibility of a soil to erosion, without regard to topography, vegetation cover, management or weather conditions.

- **soil erosion**
  the processes by which soil is removed from one place by forces such as wind, water, waves, glaciers and construction activity and eventually deposited at some new place.

- **soil fertility**
  the ability of a soil to support plant growth through

providing water, nutrients and a growth medium.

- **soil horizon**
layer of soil differing from adjacent genetically related layers in physical, chemical and biological properties or characteristics. Layer of soil, usually approximately parallel to the soil surface, with distinct characteristics produced by soil forming processes.

- **soil moisture (soil water)**
water diffused in the soil, the upper part of the zone of aeration from which water is discharged by the transpiration of plants or by soil evaporation. See Field-moisture capacity and Field-moisture deficiency.

- **soil moisture (water) depletion (deficit)**
difference between field capacity and the actual soil moisture in the root zone soil at any given time. It is the amount of water required to bring the soil in the root zone to field capacity. Amount of water required to fill the plant root zone to field capacity.

- **soil profile**
vertical section of the soil from the surface through all its horizons into the parent material. A vertical cross section through a soil, which shows any layers or horizons of which the soil is composed.

- **soil reservoir capacity**
preferred term plant available water.

- **soil structure**
the way in which individual soil particles form aggregates, particularly the shapes and arrangement of such aggregates especially important to soil hydrologic characteristics.

- **soil subsidence**
lowering of the normal level of the ground, usually due to over pumping of water or oil from wells.

- **soil texture**
the mix of different sizes of particles in a soil.

- **soil water (moisture)**
all water stored in the soil.

- **soil water characteristic curve**
soil-specific relationship between the soil-water matric

potential and soil-water content.

- **soil water content**
  amount of water in a given volume (or weight) of soil. Amount of water in a soil sample based on dry weight of the soil sample.

- **soil-water (moisture) deficit or depletion**
  depth of water required to bring a specific depth of soil to field capacity at a particular time. Preferred term soil moisture depletion.

- **soil-water tension**
  measure of the tenacity with which water is retained in the soil. It is the force per unit area that must be exerted to remove water from the soil.

- **solenoid**
  electromagnet which is connected to a controller and facilitates the opening and closing of automatic control valves (or valve-in-head sprinklers).

- **solid set irrigation system**
  see **irrigation system.**

- **solid waste**
  refuse materials composed primarily of solids at normal ambient temperatures.

- **solute**
  any substance derived from the atmosphere, vegetation, soil or rock that is dissolved in water.

- **solution**
  a mixture of a solvent and a solute. In some solutions, such as sugar water, the substances mix so thoroughly that the solute cannot be seen. But in other solutions, such as water mixed with dye, the solution is visibly changed.

- **solvent**
  a substance that dissolves other substances, thus forming a solution. Water dissolves more substances than any other and is known as the 'universal solvent'.

- **somatic damage**
  nonhereditary damage to individual cell tissues from radiation.

- **SP**
  abbreviation for an unknown species.

- **spacing**
  between sprinklers (S_s, S) distance from one sprinkler to the next one along a row.

between rows (of sprinklers) (S_r) distance between adjacent rows of sprinklers.
between drains (S_d) distance between parallel subsurface drain lines.
equilateral triangular spacing(S) sprinklers are spaced in an equilateral triangular pattern.
rectangular spacing Sprinklers are spaced in a rectangular pattern.
row spacing
distance between the rows of sprinklers.
head to head spacing (S) sprinklers are spaced so that each head throws water to the adjacent heads.
triangular spacing (S) sprinklers are spaced in a triangular pattern.

- **species**

a group of plants having common characteristics, distinct from others of the same genus. The basic unit in classification. A group of plants that will propagate among themselves but not normally with members of another group and will breed true.

- **specific conductance**

a measure of the ability of a water to conduct an electrical current. Specific conductance is related to the type and concentration of ions in solution and can be used for approximating the dissolved solids concentration in water.

- **specific conductance**

a measure of the ability of water to conduct an electrical current as measured using a 1-cm cell and expressed in units of electrical conductance, i.e., Siemens per centimetre at 25 degrees celsius. Specific conductance can be used for approximating the total dissolved solids content of water by testing its capacity to carry an electrical current. In water quality, specific conductance is used in ground water monitoring as an indication of the presence of ions of chemical substances that may have been released by a leaking landfill or other waste storage or disposal facility. A higher specific conductance in water drawn from down gradient wells when compared to up gradient wells indicates possible contamination from the facility.

- **specific gravity**
  ratio of a substance's density (or specific weight) to that of some standard substance such as water. For liquids, the standard is water at sea level and 60 °F (SP. GR. = 1).

- **specific heat**
  the amount of heat required to raise the temperature of a kilogram of a substance (water) by 1 degree celsius.

- **specific speed**
  index of pump type related to impeller speed, discharge and total head.

- **specific volume**
  volume occupied by a unit mass of fluid.

- **specific weight**
  weight per unit volume of a substance.

- **specimen**
  a superb or unique plant, a plant with special qualities that would warrant its use individually in such a way to display those qualities.

- **spent fuel**
  nuclear material that is no longer capable of sustaining the fission process.

- **sphagnum peat**
  commonly sold under the name 'Peat Moss'.

- **spigot**
  a spigot is a male connection on a fitting. A spigot fits inside a socket. It can be threaded or glued, but most of the time the term is used for glued fittings.

- **spike**
  a grouping of flowers consisting of a central stalk with flowers directly attached.

- **spile**
  conduit, made of lath, pipe or hose, placed through

ditch banks to transfer water from an irrigation ditch to a field.

- **spillway**
  the channel or passageway around or over a dam through which excess water is diverted.

- **spittle bug**
  young nymphs cover themselves with a white frothy mass of 'spittle' and may relocate on a plant several times before reaching adulthood. Adults resemble leafhoppers. Both feed on the sap of twigs and foliage.

- **spray irrigation**
  an common irrigation method where water is shot from high-pressure sprayers onto crops. Because water is shot high into the air onto crops, some water is lost to evaporation.

- **spring**
  1. a place where the water flows from the ground. In most cases springs are the points at which the underground water table intercepts the earth's surface, and thus, water flows from the ground. Some springs are the outlets of underground streams or rivers that flow through channels in the ground.
  2. the coiled metal device that retracts a sprinkler head or helps a hydraulic valve close.

- **sprinkler**
  sprinkler heads are devices which distribute water over a given area for irrigation (or to put out fires). The primary purpose of sprinklers, however, is to get golfers wet on cold mornings.

- **sprinkler (head)**
  1. device for distributing water under pressure.

Nozzle or device, which may or may not rotate, for distributing water under pressure through the air.
2. hydraulically operated mechanical device which discharges water through a nozzle or nozzles.
The various types of sprinklers are:
Gear drive sprinkler: Sprinkler containing gears as part of its rotational drive mechanism.
Impact drive: Sprinkler which rotates using a weighted or spring-loaded arm which is propelled by the water stream and hits the sprinkler body, causing movement.
Rotor: Sprinkler that rotates, but may more specifically refer to a gear driven sprinkler.
Spray head: Sprinkler head that does not rotate.
Valve-in-head: Sprinkler head having an integrally mounted valve.

- **sprinkler distribution pattern**
water depth-distance relationship measured from a single sprinkler head. Two dimensional water depth-distance relationship measured from a single (or multiple) sprinkler head(s).

- **sprinkler irrigation**
a method of irrigation in which the water is sprayed, or sprinkled, through the air to the ground surface.

- **sprinkler irrigation system**
see **irrigation system.**

- **sprinkler precipitation rate**
see **precipitation rate**.

- **sprinkler profile**
see **profile (sprinkler).**

- **square spacing**
the term given to a sprinkler head layout pattern where the sprinklers, when viewed from above, appear as more or less a square with one sprinkler in each corner. See also the more efficient 'triangle spacing'. Square spacing is the preferred sprinkler layout of non-hip people.

- **stage**
the height of a water surface above an established datum plane; also gauge height.

- **stage, flood**
see **flood stage**

- **stage-capacity curve**
  a graph showing the relation between the surface elevation of the water in a reservoir, usually plotted as ordinate, against the volume below that elevation, plotted as abscissa.

- **stage-discharge curve (rating curve).**
  a graph showing the relation between the gage height, usually plotted as ordinate, and the amount of water flowing in a channel, expressed as volume per unit of time, plotted as abscissa.

- **stage-discharge relation**
  the relationship expressed by the stage-discharge curve.

- **standard**
  generally, a single, upright stem topped by a rounded crown of foliage.

- **standard deviation (SD)**
  Statistical term used to describe the distribution of values.

- **standard dimension ratio (SDR)**
  dimension ratio is defined as the ratio of the diameter of a pipe to its wall's thickness. Outside diameter is used for OD rated pipe while ID is used for ID rated pipe. Certain dimension ratios have been selected by convention and standards to be used for construction of pipe. These dimension ratios are referred to as standard dimension ratios.

- **standard project flood (SPF)**
  a hypothetical flood resulting from the most severe combination of meteorological and hydrological conditions that are reasonably characteristic of a region. It is comparable to historical great floods which have been reported.

- **static lift**
  vertical distance between water source and discharge water levels in a pump installation. Same as total static head.

- **static pressure**
  pressure in a closed system, without any water movement. Water pressure at the time when the water is at rest.

- **static water level**
  depth to the water surface in a well under static conditions (not being pumped). May be subject to seasonal

changes or lowering due to depletion.

- **static water pressure**
the water pressure as measured when the water is not moving. The 'not moving' part is critical, if the water is moving it isn't 'static'. When measuring static water pressure all the water outlets on the pipe must be closed. So if you're measuring the static pressure at a house you connect the pressure gauge, then take the reading while all the faucets, the ice maker, etc., are turned off. Static water pressure is a measure of the water's energy potential.

- **station**
circuit on a controller which has the ability to be programmed with a run time unique and separate from other circuits and provides power to one or more remote control valves (or valve-in-head sprinklers).

- **stationary sprinklers**
irrigation having underground piping with risers and sprinklers.

- **stem flow**
rainfall or snowmelt led to the ground down the trunks or stems of plants.

- **stockpiling**
amassing amounts of some substance well beyond present needs in anticipation of a shortage of that substance.

- **stolon**
1. a stem that creeps along the ground
2. the tips taking root and forming a new plant.

- **storage**
1. water artificially impounded in surface or underground reservoirs, for future use. The term regulation refers to the action of this storage in modifying streamflow. See also '**Conservation storage**, **Total storage**, **Dead storage**, and **Usable storage**'.
2. water naturally detained in a drainage basin, such as ground water, channel storage, and depression storage. The term 'drainage basin storage' or simply 'basin storage' is sometimes used to refer collectively to the amount of

water in natural storage in a drainage basin.

- **storage ratio**
the net available storage divided by the mean flow for 1 year.

- **storage, bank**
see **bank storage**.

- **storage, conservation**
see **conservation storage**.

- **storage, dead**
see **dead storage**.

- **storage, depression.**
see **depression storage**.

- **storage, total**
see **total storage**.

- **storage-required frequency curve**
a graph showing the frequency with which storage equals or exceeds the selected amounts, will be required to maintain selected rates of regulated flow.

- **storm**
a disturbance of the ordinary average conditions of the atmosphere which, unless specifically qualified, may include any or all meteorological disturbances, such as wind, rain, snow, hail, or thunder.

- **storm seepage**
that part of precipitation which infiltrates the surface soil, and moves toward the streams as ephemeral, shallow, perched ground water above the main ground-water level. Storm seepage is usually part of the direct runoff.

- **storm sewer**
a sewer that carries only surface runoff, street wash, and snow melt from the land. In a separate sewer system, storm sewers are completely separate from those that carry domestic and commercial wastewater (sanitary sewers).

- **storm water**
rainfall that does not percolate into the ground or evaporate.

- **storm flow**
see **direct runoff**.

- **storm water discharge**
precipitation that does not infiltrate into the ground or evaporate due to impervious land surfaces but instead flows onto adjacent land or water areas and is routed into drain/sewer systems.

- **stratification**
  a layering of a water body caused by differences in water density. It is commonly caused by temperature or salinity differences.

- **stratified estuary**
  an arm of the sea in which fresh water from the land overlies denser salt water.

- **stratosphere**
  layer of the atmosphere between 3 and 30 mi (5 and 50 km) in altitude, characterised by increasing temperature with altitude.

- **stream**
  a general term for a body of flowing water. In hydrology the term is generally applied to the water flowing in a natural channel as distinct from a canal. More generally as in the term stream gagging, it is applied to the water flowing in any channel, natural or artificial.

- **stream gauging**
  the process and art of measuring the depths, areas, velocities, and rates of flow in natural or artificial channels.

- **stream order**
  a method of numbering streams as part of a drainage basin network. The smallest entrenched mapped tributary is called first order, the stream receiving the tributary is called second order, and so on. It is usually necessary to specify the scale of the map used. Tributaries which have no branches are designated as of the first order, streams which receive only first-order tributaries are of the second order, larger branches which receive only first-order and second-order tributaries are designated third order, and so on, the main stream being always of the highest order.

- **stream segment**
  refers to the surface waters of an approved planning area exhibiting common biological, chemical, hydrological, natural, and physical characteristics and processes. Segments will normally exhibit common reactions to external stress such as discharge or pollutants.

- **streamflow**
  the discharge that occurs in a natural channel. Although the term discharge can be applied to the flow of a canal, the word streamflow uniquely describes the discharge in a surface stream course. The term 'streamflow' is more general than runoff, as streamflow may be applied to discharge whether or not it is affected by diversion or regulation.

- **streamflow depletion**
  the amount of water that flows into a valley, or onto a particular land area, minus the water that flows out the valley or off from the particular land area.

- **stream-gauging station**
  a gagging station where a record of discharge of a stream is obtained. Within the Geological Survey this term is used only for those gagging stations where a continuous record of discharge is obtained.

- **stress**
  a condition that places a plant's health in danger. Common stresses are drought, too much water or too much heat or sun.

- **stress irrigation**
  management of irrigation water to apply less than enough water to satisfy the soil water deficiency in the entire root zone. Preferred term is limited irrigation or deficit irrigation.

- **strip cropping**
  a soil conservation technique in which parallel strips of land are planted in different crops.

- **strip-mining or surface mining**
  extraction of a mineral from the ground by excavation at the ground surface

- **structural planting**
  those plants that form the structure for outdoor spaces, such as trees, shrubs and plant masses.

- **structure (soil)**
  combination or arrangement of primary soil particles into secondary units or peds. The secondary units are characterised on the basis of size, shape, and grade (degree of distinctness). Combination of various soil particle types into a uniform mixture

that behaves as a single unit. Structural classes include granular, blocky, and columnar massive, platy, prismatic, among others. Aggregation of primary soil particles, into units which are separated from each other by surfaces of weakness. An individual natural soil aggregate is called a ped in contrast to a clod caused by disturbance, or a concretion caused by cementation.

- **stubble mulch**
  a soil covering composed of the unused stalks of crop plants.

- **sub-humid climate**
  see **climate**.

- **sub-irrigation**
  applying irrigation water below the ground surface either by raising the water table within or near the root zone, or by use of a buried perforated or porous pipe system which discharge's directly into the root zone.

- **sublimation**
  the transition of water directly from the solid state to the gaseous state, without passing through the liquid state; or vice versa. Compare condensation, evaporation.

- **sub meander**
  small meander contained with banks of main channel, associated with relatively low discharges.

- **submergence**
  applied to submersible pumps: Distance beneath the static water level, at which a pump is set. Synonym: immersion level. Total Dynamic Head
  vertical lift + friction loss in piping (see **friction loss**).

- **submersible pump**
  a motor/pump combination designed to be placed entirely below the water surface.

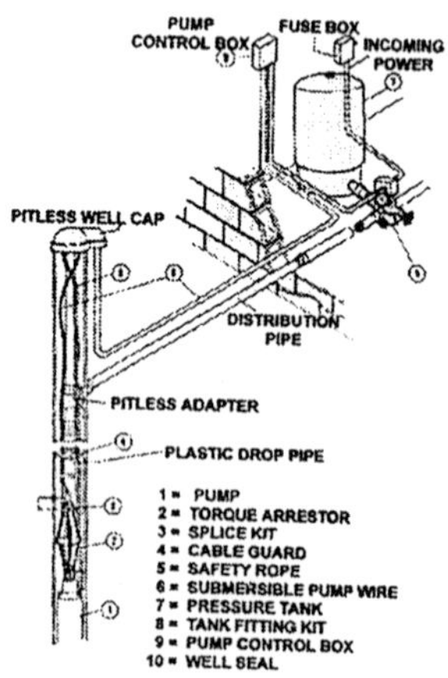

- **subsidence**
lowering of the soil level caused by the shrinkage of organic layers. This shrinkage is due to desiccation, consolidation and biological oxidation.

- **subsidence inversion**
a temperature inversion caused by differential warming of a sinking air mass. Upper portions of the mass gets warmed more than lower portions, causing the inversion.

- **substitutability**
the degree to which one material can be substituted for another in end uses.

- **subsurface drip irrigation**
application of water below the soil surface through emitters, with discharge rates generally in the same range as drip irrigation. This method of water application is different and is not to be confused with sub-irrigation where the root zone is irrigated by water table control.

- **subsurface runoff**
see **storm seepage**.

- **subsurface water**
all water, solid, liquid or gaseous, that occurs beneath the Earth's surface; located below the water table in the zone of saturation.

- **sucker**
a strong, vertical shoot growing from below the ground, often rising from the root.

- **suction (static) lift (FT)**
Vertical distance between the elevation of the surface of the water source and the centre of the pump impeller.
suction lift
applied to surface pumps: Vertical distance from the surface of the water in the source, to a pump located above surface pump located above. This distance is limited by physics to around 20 feet at sea level (subtract 1 ft. per 1000 ft. altitude) and should be minimized for best results.

- **sulphur content**
the amount of sulphur found in coal. Combustion of coal with a high sulphur content results in emissions of sulphur oxides, which contributes to acid precipitation.

- **supplemental irrigation**
commonly, irrigation as carried on in humid areas.

The term means that the irrigation water is supplementary to the natural rainfall rather than being the primary source of moisture as in the arid and semiarid West. Supplementary irrigation is used generally to prevent retardation of growth during periods of drought.

- **supplemental sources**
when irrigation water supplies are obtained from more than one source, the source furnishing the principal supply is commonly designated the primary source, and the sources furnishing the additional supplies, the supplemental sources.

- **supplemental water requirements**
application of water in addition to natural precipitation needed to maintain the optimum health and appearance of landscape plantings. See **water budget.**

- **surface impoundment**
an indented area in the land's surface, such as pit, pond, or lagoon.

- **surface irrigation**
application of water by means other than spraying such that it can contact between the edible portion of any food crop and the irrigation water is prevented.

- **surface irrigation**
see **irrigation systems.**

- **surface pump**
a pump that is not submersible. It must be placed no more than about 18 ft. above the surface of the water in the well. See **priming** and **jet pump**.

- **surface runoff**
that part of the runoff which travels over the soil surface to the nearest stream channel. It is also defined as that part of the runoff of a drainage basin that has not passed beneath the surface since precipitation. The term is misused when applied in the sense of direct runoff.

- **surface soil**
upper part of the soil ordinarily moved in tillage, or its equivalent in uncultivated soils, about 10 to 20 cm in thickness.

- **surface tension**
  the attraction of molecules to each other on a liquid's surface. Thus, a barrier is created between the air and the liquid.

- **surface water**
  water on the earth's surface. Lakes, bays, ponds, impounding reservoirs, springs, rivers, streams, creeks, estuaries, wetlands, marshes, inlets, canals, gulfs inside the territorial limits of the state, and all other bodies of surface water, natural or artificial, inland or coastal, fresh or salt, navigable or non-navigable, and including the beds and banks of all watercourses and bodies of surface water, that are wholly or partially inside or bordering the state or subject to the jurisdiction of the state; except that waters in treatment systems which are authorised by state or federal law, regulation, or permit, and which are created for the purpose of waste treatment are not considered to be waters in the state.

- **surge irrigation**
  a surface irrigation technique wherein flow is applied to furrows (or less commonly, borders) intermittently during a single irrigation set.

- **surge irrigation**
  see **irrigation systems**.

- **surge pressure (PS) (PSI)**
  water pressure caused by changes in water velocity in a pipe system.

- **suspended particulates**
  in reference to air quality, solid or liquid particles with diameters from 0.03 to 100 microns.

- **suspended sediment**
  very fine soil particles that remain in suspension in water for a considerable period of time without contact with the bottom. Such material remains in suspension due to the upward components of turbulence and currents and/or by suspension.

- **suspended solids**
  solids that are not in true solution and that can be removed by filtration. Such suspended solids usually contribute directly to

turbidity. Defined in waste management, these are small particles of solid pollutants that resist separation by conventional methods.

- **suspended-sediment concentration**
  the ratio of the mass of dry sediment in a water-sediment mixture to the mass of the water-sediment mixture. Typically expressed in milligrams of dry sediment per litre of water-sediment mixture.

- **suspended-sediment discharge**
  the quantity of suspended sediment passing a point in a stream over a specified period of time. When expressed in tons per day, it is computed by multiplying water discharge (in cubic feet per second) by the suspended-sediment concentration (in milligrams per litre) and by the factor 0.0027.

- **sustainability**
  economic growth with environmental responsibility; economic activity that could be carried on indefinitely without resource depletion.

- **sustainable agriculture**
  an agricultural system that is dependent solely on renewable resources and that maintains the soil in such a condition so that it will continue to be productive indefinitely.

- **sustainable management**
  method of exploiting a resource that can be carried on indefinitely. Removal of water from an aquifer in excess of recharge is, in the long term, not a sustainable management method.

- **sustained overdraft**
  long term withdrawal from the aquifer of more water than is being recharged.

- **sustained yield**
  management of renewable resources conducted in such a way as to allow a constant rate of harvest indefinitely.

- **swale**
  a depression in grading used to carry water; a man-made 'river' used to drain a site. To prevent erosion, a swale is often lined with rip-rap, or native grasses. A functional and ornamental swale is called a dry stream.

- **swim plan**
surface water improvement and management plan.

- **swing check**
see **valve**.

- **swing joint**
threaded connection of pipe and fittings between the pipe and sprinkler which allows movement to be taken up in the threads rather than as a sheer force on the pipe. Also used to raise or lower sprinklers to a final grade without plumbing changes.

- **synergistic impacts**
effects of a system's activity that are different from the individual effects of component parts of the system.

- **synfuel**
a contraction of synthetic fuel, liquid or gaseous fossil fuel manufactured from other fuels that are less useful as found in nature.

- **system capacity**
ability of an irrigation system to deliver the net required rate and volume of water necessary to meet crop water needs plus any losses during the application process. Crop water needs can include soil moisture storage for later plant use, leaching of toxic elements from the soil, air temperature modification, crop quality, and other plant needs.

- **system precipitation rate**
see **precipitation rate.**

- **systemic**
any chemical absorbed into a plant's system. There are systemic insecticides, fungicides and weed killers.

- **tail water**
water in a stream or canal, immediately downstream from a structure. Excess irrigation water which reaches the lower end of a field.

- **tailings**
residue of raw material or waste separated out during the processing of crops or mineral ores.

- **tailpipe standards**
emissions limitations applicable to mobile source engine exhausts.

- **tampering**
adjusting, negating, or removing pollution control equipment on a motor vehicle.

technical-grade active ingredient (TGA) A pesticide chemical in pure form as it is manufactured prior to being formulated into an end-use product (e.g., wettable powders, granules, emulsifiable concentrates). Registered manufactured products composed of such chemicals are known as technical grade products.

- **tank**
  an artificial reservoir for stock water.

- **taproot**
  the main' root growing straight down, like the root of a carrot.

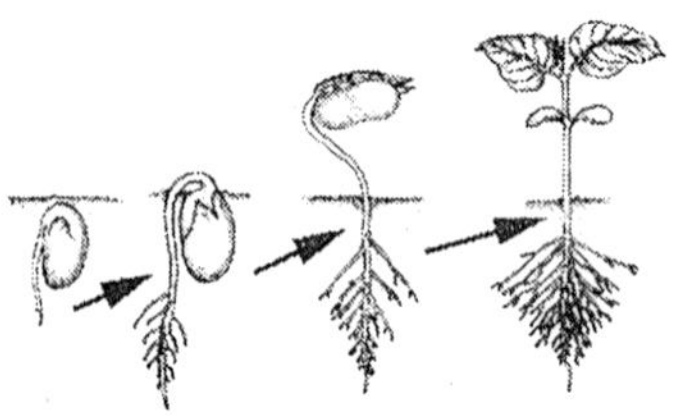

- **tar sand**
  sandy deposits containing heavy oil or tar. The sand must be heated to extract the oil.

- **traffic analysis zone (TAZ)**
  a geographical area used in transportation planning.

- **total dissolved solids (TDS)**
  the sum or all inorganic and organic particulate material. TDS is an indicator test used for wastewater analysis and is also a measure of the mineral content of bottled water and groundwater. There is a relationship between TDS and conductivity.

- **technology-based limitations**
  industry-specific effluent limitations based on best available preventive technology applied to a discharge when it will not cause a violation of water quality standards at low stream flows. Usually applied to discharges into large rivers.

- **technology-based standards**
  industry-specific effluent limitations applicable to direct and indirect sources which are developed on a category-by-category basis using statutory factors, not including water-quality effects.

- **technology-based treatment requirements**
  permit requirements based on the application of pollution treatment or control technologies including BTP (Best Practicable Technol-

ogy), BCT (Best Conventional Technology), BAT (Best Available Technology)- economically achievable, and NSPS (New Source Performance Standards).

- **tee**
  a tee is a fitting used to branch a side pipe off of a pipeline.

- **temperate forest**
  a biome characterised primarily by deciduous broad-leaved trees.

- **temperature inversion**
  a condition in the atmosphere in which warm air overlies cool air. Inversions restrict vertical air circulation.

- **tender**
  denotes a low tolerance of freezing temperatures.

- **tensiometer**
  instrument consisting of a porous cup filled with water and connected to a manometer or vacuum gage, used for measuring the soil water metric potential.

- **terminal branches/growth**
  the branches at the uppermost or outermost reaches of the plant where growth originates.

- **terrace**
  a berm or discontinuous segments of a berm, in a valley at some height above the flood plain, representing a former abandoned flood plain of the stream.

- **terracing**
  a soil and water conservation technique consisting of ridges on the contour, or level areas constructed on a slope.

- **territorial sea**
  a band of open ocean adjacent to the coast, over which the coastal nation has control. It is generally either 3 or 12 nautical miles (5.6 or 22 km) wide.

- **tertiary treatment**
  advanced cleaning of wastewater that goes beyond the secondary or biological stage, removing nutrients such as phosphorus, nitrogen, and most BOD and suspended solids.

- **tertiary wastewater treatment**
  selected biological, physical, and chemical separation processes to remove organic and inorganic substances that resist conventional treatment practices; the additional treatment of effluent beyond that of primary and secondary treatment methods to obtain a very high quality of effluent. The complete wastewater treatment process typically involves a three-phase process: First, in the primary wastewater treatment process, which incorporates physical aspects, untreated water is passed through a series of screens to remove solid wastes. Second, in the secondary wastewater treatment process, typically involving biological and chemical processes, screened wastewater is then passed a series of holding and aeration tanks and ponds; and Third, the tertiary wastewater treatment process consists of flocculation basins, clarifiers, filters, and chlorine basins or ozone or ultraviolet radiation processes.

- **textural class**
  classification used to convey an idea of the textural makeup of soils and to give an indication of their physical properties. Three broad groups of these classes are recognised – sands, loams, and clays. Within each group specified textural class names have been devised (i.e. loamy sand). Classification name given a soil because of the particular size groups of particles found in the 'A' horizon.

- **texture (soil)**
  relative proportions of sand, silt and clay in a given soil. Relative proportions of the various soil separates in a soil as described by the classes of soil texture.

- **thalweg**
  the line of maximum depth in a stream. The thalweg is the part that has the maximum velocity and causes cutbanks and channel migration.

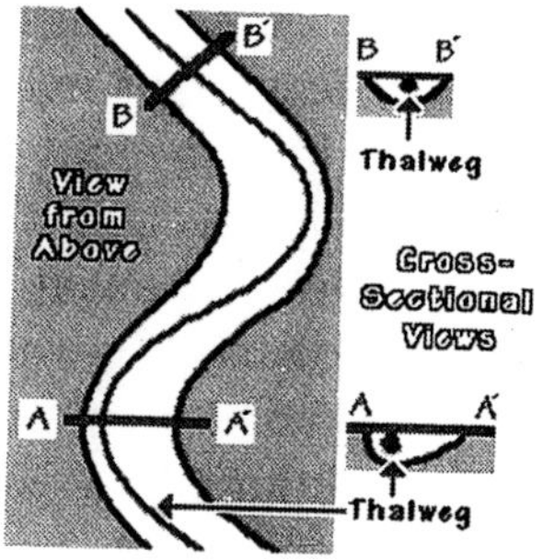

- **thatch**
building up of organic material at the base of turf grass leaf blades that repels water and reduces water infiltration capacity.

- **theoretical maximum residue contribution (TMRC)**
the theoretical maximum amount of a pesticide in the daily diet of an average person. It assumes that the diet is composed of all food items for which there are tolerance-level residues of the pesticide. The TMRC is expressed as milligrams of pesticide/kilograms of body weight/day.

- **thermal gradient**
temperature difference between two areas.

- **thermal pollution**
discharge of heated water from industrial processes that can kill or injure aquatic organisms.

- **thermal pollution**
a reduction in water quality caused by increasing its temperature, often due to disposal of waste heat from industrial or power generation processes. Thermally polluted water can harm the environment because plants and animals can have a hard time adapting to it.

- **thermal stratification**
(of a lake) vertical temperature stratification that shows the following: The upper layer of the lake, known as the epilimnion, in which the water temperature is virtually uniform; a stratum next below, known as the thermocline, in which there is a marked drop in temperature per unit of depth; and the lowermost region or stratum, known as the hypolimnion, in which the temperature from its upper limit to the bottom is nearly uniform.

- **thermal treatment**
use of elevated temperatures to treat hazardous wastes.

- **thermocline**
  a zone in a water body in which temperature declines rapidly with increasing depth. Vertical water circulation is limited by the presence of a thermocline.

- **thermoelectric power water use**
  water used in the process of the generation of thermoelectric power. Power plants that burn coal and oil are examples of thermoelectric-power facilities.

- **threatened species**
  a species that is not endangered but has a rapidly declining population.

- **threshold**
  the lowest dose of a chemical at which a specified measurable effect is observed and below which it is not observed.

- **threshold level**
  time-weighted average pollutant concentration values, exposure beyond which is likely to adversely affect human health.

- **threshold limit value (TLV)**
  the concentration of an airborne substance to which an average person can be repeatedly exposed without adverse effects. TLVs may be expressed in three ways:
  TLV-TWA: time weighted average, based on an allowable exposure averaged over a normal 8-hour work-day or 40-hour work- week
  TLV-STEL: short-term exposure limit or maximum concentration for a brief specified period of time, depending on a specific chemical (TWA must still be met) and
  TLV-C: ceiling Exposure Limit or maximum exposure concentration not to be exceeded under any circumstances. (TWA must still be met.)

- **threshold pollutant**
  substance that is harmful to a particular organism only above a certain concentration, or threshold level.

- **through fall**
  in a vegetated area, the precipitation that falls directly to the ground or the rainwater or snowmelt that drops from twigs or leaves.

- **thrust block**
  normally, concrete poured in place at changes in direc-

tion of water flow in piping systems to prevent movement of the pipe.

- **tidal marsh**
  low, flat marshlands traversed by channels and tidal hollows, subject to tidal inundation; normally, the only vegetation present is salt-tolerant bushes and grasses.

- **tidal power**
  energy generated by using tidal water-level differences to drive a turbine.

- **tillage**
  ploughing, seedbed preparation, and cultivation practices.

- **time-weighted average (TWA)**
  In air sampling, the average air concentration of contaminants during a given period.

- **tilth**
  physical condition of the soil in relation to plant growth.

- **time of concentration**
  the time required for water to flow from the farthest point on the watershed to the gauging station.

- **timer/controller**
  a device operating in conjunction with electric valves to automatically control the drip irrigation system. Can set watering days of the week, run time per station, and length of time to run each station. Battery operated valves are also available for hose bibb connections.

- **tolerance petition**
  a formal request to establish a new tolerance or modify an existing one.

- **tolerances**
  permissible residue levels for pesticides in raw agricultural produce and processed foods. When-ever a pesticide is registered for use on a food or a feed crop, a tolerance (or exemption from the tolerance requirement) must be established.

- **tonnage**
  the amount of waste that a landfill accepts, usually expressed in tons per month. The rate at which a landfill accepts waste is limited by the landfill's permit.

- **top dress**
  to apply on the surface, usually referring to spreading organic material on the surface of the soil.

- **topography**
  the physical features of a surface area including relative elevations and the position of natural and man-made (anthropogenic) features.

- **topsoil**
  soil taken from the upper part of a site. Colorado Springs has a minimal amount of topsoil. It is more dominant in the Midwest and less-arid areas.

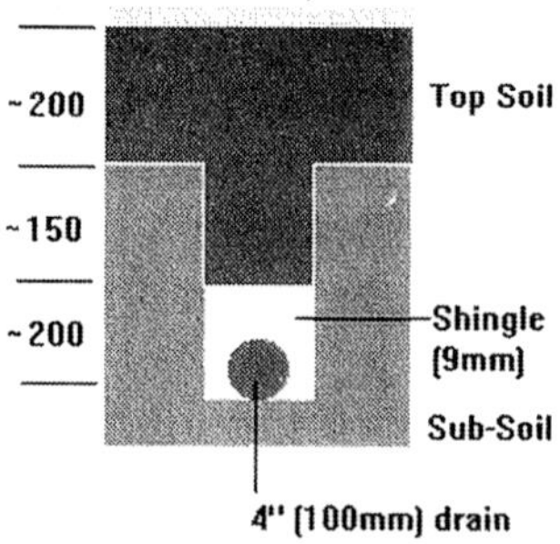

TOP SOIL FILLED

- **total dissolved phosphorous**
  the total phosphorous content of all material that will pass through a filter, which is determined as orthophosphate without prior digestion or hydrolysis. Also called soluble P or ortho P.

- **total dissolved solids (TDS)**
  All material that passes the standard glass river filter are now called total filterable residue. Term is used to reflect salinity.

- **total dissolved solids (TDS)**
  the total amount of dissolved solid matter found in a sample of water.

- **total dynamic head**
  see **head.**

- **total pressure head**
  the sum of all the factors which increase or decrease the available water pressure.

- **total recovered petroleum hydrocarbon**
  a method for measuring petroleum hydrocarbons in samples of soil or water.

- **total static head**
  see **static lift.**

- **total storage**
  the volume of a reservoir below the maximum controllable level including dead storage.

- **total suction head**
  see **head.**

- **total suspended solids (TSS)**
  a measure of the suspended solids in wastewater, effluent, or water bodies, determined by tests for 'total suspended non-filterable solids'.

• **total suspended particles (TSP)**
a method of monitoring airborne particulate matter by total weight.

• **toxaphene**
chemical that causes adverse health effects in domestic water supplies and is toxic to fresh water and marine aquatic life.

• **toxic chemical use substitution**
replacing toxic chemicals with less harmful chemicals in industrial processes.

• **toxic cloud**
airborne plume of gases, vapours, fumes, or aerosols containing toxic materials.

• **toxic concentration**
the concentration at which a substance produces a toxic effect.

• **toxic dose**
the dose level at which a substance produces a toxic effect.

• **toxic pollutants**
materials that cause death, disease, or birth defects in organisms that ingest or absorb them. The quantities and exposures necessary to cause these effects can vary widely.

• **toxic substance**
a chemical or mixture that may present an unreasonable risk of injury to health or the environment.

• **toxic waste**
a waste that can produce injury if inhaled, swallowed, or absorbed through the skin.

• **toxicant**
a harmful substance or agent that may injure an exposed organism.

• **toxicity**
the degree to which a substance or mixture of substances can harm humans or animals. Types of Toxicity are as follows:

1. acute toxicity involves harmful effects in an organism through a single or short-term exposure.
2. chronic toxicity is the ability of a substance or mixture of substances to cause harmful effects over an extended period, usually upon repeated or continuous exposure sometimes lasting for the entire life of the exposed organism.
3. sub-chronic toxicity is the ability of the substance to cause effects for more than one year but less than the lifetime of the exposed organism.

- **toxicity assessment**
characterisation of the toxicological properties and effects of a chemical, with special emphasis on establishment of dose-response characteristics.

- **toxicity reduction evaluation (TRE)**
a study conducted to determine the source(s) of toxicity in a discharge effluent so that these sources can be controlled sufficiently to allow a discharger to comply with their permit limits.

- **toxicity test**
the means to determine the toxicity of a chemical or an effluent using living organisms. A toxicity test measures the degree of response of an exposed test organism to a specified chemical or effluent.

- **toxicity testing**
biological testing (usually with an invertebrate, fish, or small mammal) to determine the adverse effects of a compound or effluent.

- **toxicological profile**
an examination, summary, and interpretation of a hazardous substance to determine levels of exposure and associated health effects.

- **tracker**
a mounting rack for a PV array that automatically tilts to follow the daily path of the sun through the sky. A 'tracking array' will produce more energy through the course of the day, than a 'fixed array'(non-tracking) particularly during the long days of summer. Energy gain varies from 10-25%. When used with a Maximum

Power Point Tracker, total yearly gains typically approach 30%.

- **tragedy of the commons**
the idea that no one takes responsibility for things that everybody owns.

- **trajectory**
angle above horizontal of the stream of water as it leaves the sprinkler head.

- **trans-boundary pollutants**
air pollution that travels from one jurisdiction to another, often crossing state or international boundaries. Also applies to water pollution.

- **trans-boundary pollution**
transport of pollutants (particularly air pollutants) across national or state boundaries.

- **transfer station**
facility where solid waste is transferred from collection vehicles to larger trucks or rail cars for longer distance transport.

- **transformer**
an electrical device that steps up voltage and steps down current proportionally (or vice-versa). Transformers work with AC only. (For DC, see **Converter**). Mechanical analogy, gears or belt drive.

- **transient water system**
a non-community water system that does not serve 25 of the same non-residents per day for more than six months per year.

- **translocation**
movement of water to other areas than where it was applied.

- **transmissibility**
the capacity of a rock to transmit water under pressure. The coefficient of transmissibility is the rate of flow of water, at the prevailing water temperature, in gallons per day, through a vertical strip of the aquifer one foot wide, extending the full saturated height of the aquifer under a hydraulic gradient of 100-percent. A hydraulic gradient of 100-percent means a one foot drop in head in one foot of flow distance.

- **transmission lines**
pipelines that transport raw water from its source to a water treatment plant, then to the distribution grid system.

- **transmissivity**
 refers to the rate at which limestone allows the transmission of water. Limestone can be highly porous, but not very transmissive if the pores are not connected to each other. Technically speaking, it is the rate at which water is transmitted through a unit width of aquifer under unit hydraulic gradient. Transmissivity is directly proportional to aquifer thickness, thus it is high where the Edwards is thick and low where it is thin, given the same hydraulic conductivity.

- **transpiration**
 process of plant water uptake and use, beginning with absorption through the roots and ending with transpiration at the leaf surfaces(in./day). (See also **evapotranspiration**) Liquid movement of water from the soil, into the roots, up the plant stems, and finally out of the plant leaves into the air as vapour.

- **transplant shock and recovery**
 the adverse impact of digging, relocating and replanting, including plants from a planting container or balled and burlap stock, on a plant's overall health and the length of time until it resumes normal growth. Although smaller plant material does not provide immediate impact in the new landscape, it is less affected by relocation and resumes normal growth much sooner than larger plant material. (See **establishment period**).

- **trash-to-energy plan**
 burning trash to produce energy.

- **traveller (travelling gun) irrigation**
 see **irrigation system.**

- **traveller head**
 atmospheric pressure, value of atmospheric pressure at a specific location and condition. The various kinds of heads are:
 head dynamic: specific energy in a flow system.
 head discharge $[H_d]$: pressure head at the outlet of the pump.
 head elevation* (H_z, Z): Head as a result of elevation above a defined datum.

head friction *(H_f): energy head loss caused by the friction of water flowing through a pipe.

head loss *(H_L) ***energy*** loss in fluid flow: **Head loss, converging and diverging pipes (H_g)(ft.)** energy loss associated with water flowing through converging or diverging pipe sections.

head loss, entrance **(H_e)(ft)**: energy loss associated with water flowing through the entrance of a pipe.

head loss, bends **(H_b)(ft)**: energy loss associated with water flowing through a bend in a pipe.

head pressure *(H_{pr}): pressure energy in a liquid system expresses as the equivalent height of a water column above a given datum.

head static* **(H_s)**: energy associated with a static liquid system composed of elevation and pressure components.

head static **discharge***: static energy components at the discharge of a pump including elevation and pressure.

head static **suction***: vertical distance from the pump centreline to the surface of the liquid when the liquid supply is above the pump.

head total (dynamic)* **(H_t, H, TDH)**: head required to pump water from its source to the point of discharge. Equal to the static lift plus friction head losses in pipes and fittings plus velocity head.

Energy in the liquid system expressed as the equivalent height of a water column above a given datum.

Sum of static, pressure, friction and velocity head that a pump works against while pumping at a specific flow rate.

head total suction*: head required to lift water from the water source to the centerline of the pump plus velocity head, entrance losses and friction losses in suction pipeline.

head vapour pressure* **(H_{vp})**: pressure head at which the liquid (water) will vaporise or boil at a given temperature.

head velocity* **(H_v)**: head or energy caused by the velocity of a moving fluid.

Amount of pressure required for generating a specific velocity. This is the amount of energy or pressure that is used to make the water move at a given velocity.

- **treated wastewater**

wastewater that has been subjected to one or more physical, chemical, and biological processes to reduce its potential of being health hazard.

- **treatment**

1. any method, technique, or process designed to remove solids and/or pollutants from solid waste, waste-streams, effluents, and air emissions.
2. methods used to change the biological character or composition of any regulated medical waste so as to substantially reduce or eliminate its potential for causing disease.

- **treatment plant**

a structure built to treat wastewater before discharging it into the environment.

treatment, storage, and disposal facility:

site where a hazardous substance is treated, stored, or disposed off.

- **tree spade**

specialised motorised equipment to dig and usually transport trees. Tree spades come in a variety of sizes appropriate to different sizes and species of trees. Generally deciduous trees require a nine inches root ball for each inch of trunk caliper; evergreens require an eight inches root ball for each inch of trunk caliper.

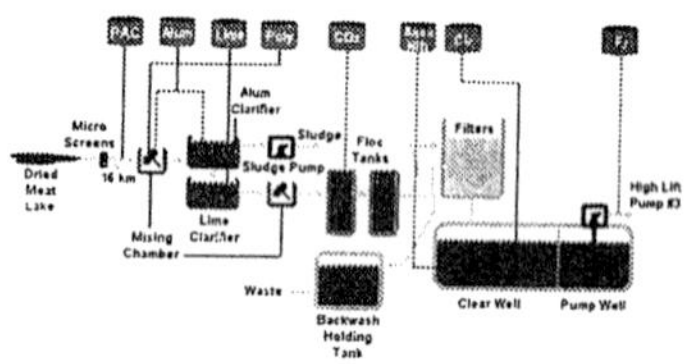

The City Of Camrose - Water Treatment Plant Schematic

- **tremie**

device used to place concrete or grout under water.

- **trend**

a statistical term referring to the direction or rate of increase or decrease in magnitude of the individual members of a time series of data when random fluctuations of individual members are disregarded.

- **triangle spacing**
  the term given to a sprinkler head layout pattern where the sprinklers, when viewed from above, appear as a more or less equilateral triangle with one sprinkler in each corner. Triangular spacing results in the most uniform and efficient water application using sprinklers. Cool designers use triangular spacing. Not so cool designers use the other type.

- **tributary**
  a smaller river or stream that flows into a larger river or stream. Usually, a number of smaller tributaries merge to form a river.

- **trickle irrigation**
  a method of micro-irrigation wherein water is applied to the soil surface as drops or small streams through emitters. (Preferred term is drip irrigation.) For a listing of related terms, please see **trickle irrigation**.

- **trickling filter**
  a coarse treatment system in which wastewater is trickled over a bed of stones or other material covered with bacteria that break down the organic waste and produce clean water.

- **trihalomethane (THM)**
  One of a family of organic compounds named as derivative of methane. THMs are generally by-products of chlorination of drinking water that contains organic material.

- **trophic level**
  one of the steps in a food chain.

- **tropical rainforest**
  a biome composed primarily of evergreen broad-leaved trees growing in tropical areas of high rainfall throughout most of the year.

- **troposphere**
  the layer of atmosphere closest to the Earth, extending seven to ten miles above the surface, containing most of the clouds and moisture.

- **tube settler**
  device using bundles of tubes to let solids in water settle to the bottom for removal by conventional sludge collection means; sometimes used in sedi-

mentation basins and clarifiers to improve particle removal.

- **tuber**
  swollen underground stem. A potato.

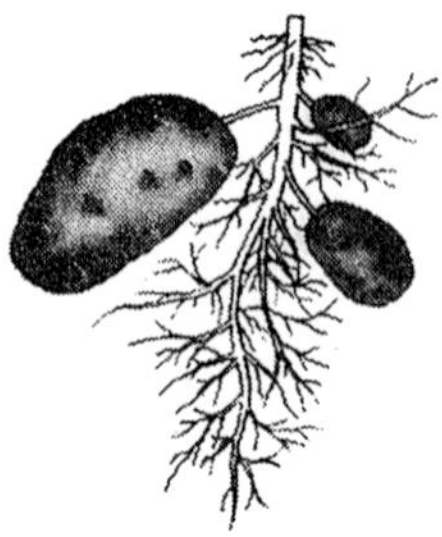

- **tuberculation**
  development or formation of small mounds of corrosion products on the inside of iron pipe. These tubercules roughen the inside of the pipe, increasing its resistance to water flow.
- **tuberous roots**
  thickened underground food storage structure. Actually a root rather than a true tuber. The dahlia is an example.
- **tundra**
  a type of treeless ecosystem dominated by lichens, mosses, grasses, and woody plants. Tundra is found at high latitudes (arctic tundra) and high altitudes (alpine tundra). Arctic tundra is underlain by permafrost and is usually water saturated.
- **turbid**
  thick or opaque with matter in suspension. Rivers and lakes may become turbid after a rainfall.
- **turbidimeter**
  a device that measures the cloudiness of suspended solids in a liquid; a measure of the quantity of suspended solids.
- **turbidity**
  1. haziness in air caused by the presence of particles and pollutants.
  2. a cloudy condition in water due to suspended silt or organic matter.
- **turbine pump**
  see **pump, vertical turbine.**

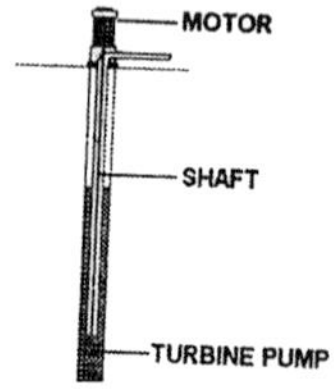

- **turbulent flow**
  flow in which the fluid particles move in an irregular random manner, in which the head loss is approxi-

mately proportional to the second power of velocity.

- **turfgrass**
  grasses that, when regularly mowed, form a dense growth of leaf blades and roots.

- **turgid**
  state of a plant cell when the cell wall is rigid due to the hydrostatic pressure of liquid in the cell.

- **unavailable soil water**
  portion of water in a soil held so tightly by adhesion and other soil forces that it cannot be absorbed by plants rapidly enough to sustain growth.

- **unconfined aquifer**
  an aquifer that discharges and recharges with an upper surface that is the water table.

- **unconsolidated formations**
  naturally occurring earth formations that have not been lithified. Alluvium, soil, gravel, clay, and overburden are some of the terms used to describe this type of formation.

- **undercurrent**
  a current below the upper currents or surface of a fluid body.

- **under-drain**
  a concealed drain with openings through which the water enters when the water table reaches the level of the drain.

- **underflow**
  the downstream flow of water through the permeable deposits that underlie a stream and that are more or less limited by rocks of low permeability.

- **under-planting**
  planting one plant beneath another, as in ground cover beneath a tree.

- **undertow**
  the current beneath the surface that sets seaward or along the beach when waves are breaking on the shore.

- **underwater**
  under the surface of the water; lying, growing, performed, worn, or operating below the surface of the water, as underwater caverns, underwater operation of a submarine.

- **uniformity**
  term describing how evenly water is applied by overlap-

ping sprinklers. Evenness of precipitation over a given area.

- **uniformity coefficient irrigation**
  characteristic of the area distribution of water in a field as the result of an irrigation.

- **uniformity coefficient (Christiansen's) (CU, CCU)**
  measure of the uniformity of irrigation water application. The average depth of irrigation water infiltrated minus the average absolute deviation from this depth, all divided by the average depth infiltrated.

- **union**
  pipe fitting used to connect two lengths of pipe in such a way that neither has to be rotated.

- **unit hydrograph**
  the hydrograph of direct runoff from a storm uniformly distributed over the drainage basin during a specified unit of time; the hydrograph is reduced in vertical scale to correspond to a volume of runoff by 1 inch from the drainage basin. The hydrograph of surface runoff (not including ground-water runoff) on a given basin due to an effective rain falling for a unit of time.

- **universal soil loss equation**
  a statistical technique for predicting the average erosion rate by rainfall under a variety of climatic, soil, topographic, and management conditions.

- **unreasonable uses**
  see **uses**.

- **unsaturated zone**
  the zone immediately below the land surface where the pores contain both water and air, but are not totally saturated with water. These zones differ from an aquifer, where the pores are saturated with water.

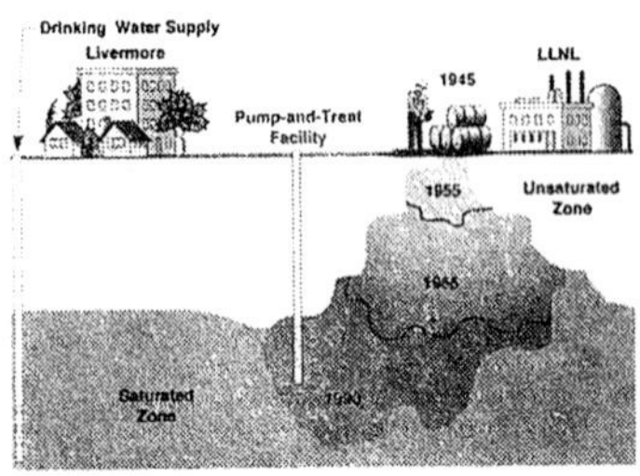

- **up-flow**
  an upward flow.

- **uplands**
  elevated areas which are characterised by non-saturated soil conditions and support flatwood vegetation.

- **upwelling**
  an upward movement of seawater that usually occurs near the margins of oceans.

- **urban runoff**
  runoff derived from urban areas, usually containing relatively high concentrations of pollutants; also called urban storm water.

- **usable precipitation**
  see **effective precipitation**.

- **usable storage**
  the volume normally available for release from a reservoir below the stage of the maximum controllable level.

- **usable storage capacity**
  quantity of groundwater of acceptable quality that can be economically withdrawn from storage.

- **uses**
  consumptive : total amount of water taken up by vegetation for transpiration or building of plant tissue, plus the unavoidable evaporation of soil moisture, snow, and intercepted precipitation associated with vegetal growth.
  non-consumptive : water that leaves the selected region and not considered consumptive. Examples are runoff, deep percolation, and canal spills.
  ***beneficial* (BU)**: beneficial use of water supports the production of crops: food, fibre, oil, landscape, turf, ornamentals, or forage.
  non-beneficial: water utilized in plant growth, which cannot be attributed as beneficial.
  reasonable : in the context of irrigation performance, all beneficial uses are considered to be reasonable uses. Non-beneficial uses are considered to be reasonable if they are justified under the particular conditions at a particular time and place.
  unreasonable : unreasonable uses are non-beneficial uses that, furthermore, are not reasonable; that is, they are without economic, practical, or other justification.

- **volts alternating current (VAC)**
most electric control valves operate on 24 VAC. That's 'alternating current', like household electricity.

- **valve**
a device used to control the flow of water. Isolation valves are used to shut-off water for repairs. Control valves turn on and off the water to the individual circuits of sprinklers or drip emitters. Check valves allow the water to flow in only one direction. Master valves are located at the water source and turn on and off the water for the entire irrigation system when not in use.

- **valve zone**
an area where the irrigation is all controlled by a single control valve. Each valve zone must be within only one hydro-zone.

- **valve-in-head sprinkler**
see **sprinkler head.**

- **vane pump**
(rotary vane) a positive displacement mechanism used in low volume high lift surface pumps and booster pumps. Durable and efficient, but requires cleanly filtered water due to its mechanical precision.

- **vapour pressure (head)**
see **head**.

- **var.**
the abbreviation for a variety. See **Variety.**

- **velocity**
usually refers to the average velocity computed as flow rate per unit area of a pipe. Is the speed at which water moves through the system (pipe).

- **velocity head**
see **head.**

- **vertical lift**
the vertical distance that water is pumped. This determines the pressure that the pump pushes against. Total vertical lift = vertical lift from surface of water source up to the discharge in the tank + discharge pres-

sure (in a pressure system). Synonym: static head. Note: Horizontal distance does NOT add to the vertical lift, except in terms of pipe friction loss. NOR does the volume (weight) of water contained in pipe or tank. Submergence of the pump does NOT add to the vertical lift in the case of a centrifugal type pump. In the case of a positive displacement pump, it may add to the lift somewhat.

- **vertical turbine pump**
  see **pump**.

- **vested water right**
  the right granted by a state water agency to use either surface or ground water.

- **virgin flow**
  the stream flow which exists or would exist if man had not modified the conditions on or along the stream or in the drainage basin.

- **visual blight**
  modification of a landscape that is visually undesirable.

- **void**
  the pore space or other openings in rock. The openings can be very small to cave size and are filled with water below the water table.

- **voids ratio**
  ratio of the volume of voids (pores) to the volume of soil.

- **voltage drop**
  loss of voltage (electrical pressure) caused by the resistance in wire and electrical devices. Proper wire sizing will minimise voltage drop, particularly over long distances. Voltage drop is determined by 4 factors: wire size, current (amps), voltage, and length of wire. It is determined by a consulting wire sizing chart or formula available in various reference tests. It is expressed as a percentage. Water analogy: friction loss in pipe.

- **voltage, open circuit**
  the voltage of a PV module or array with no load (when it is disconnected). A '12 volt nominal' PV module will produce about 18-20 volts open circuit. Abbreviation: Voc. Note that some of the so-called 'self regulating' modules produce much lower voltages and are not generally suited for water pumping.

- **voltage, peak power point**
  the voltage at which a photovoltaic module or array transfers the greatest amount of power (watts). A '12 volt nominal' PV module will typically have a peak power voltage of around 17 volts. A PV array-direct solar pump should reach this voltage in full sun conditions. In a higher voltage array, it will be a multiple of this voltage. Abbreviation: Vpp. See also **maximum power point tracker**.

- **volute**
  refers to the flow path of water and its associated pump casing as it leaves the impeller of a pump.

- **vortex emitter**
  employs a vortex effect to dissipate pressure.

- **warm season grass**
  grass that grows vigorously in warm summer months and generally loses its green colour and is dormant in the winter if the average air temperature drops below 50 to 60 degrees Fahrenheit. Some warm season grasses may die if exposed to subfreezing temperatures for extended periods.

- **wastewater**
  water containing waste including grey water, black water or water contaminated by waste contact, including process-generated and contaminated rainfall runoff.

- **wastewater reclamation**
  any process in which wastewater is put to use, such as for cooling or irrigation, with or without treatment.

- **wastewater-treatment return flow**
  water returned to the environment by wastewater-treatment facilities.

- **water**
  the liquid that descends from the clouds as rain; forms streams, lakes, and seas, and is a major constituent of all living matter. It is an odourless, tasteless, colourless, very slightly compressible liquid.

- **water allotment (WA)**
  a method to accurately and fairly estimate a total volume of water that should be allocated to a site.

- **water allotment adjustment factor(Kwa)**
  factor used in the equation to predict Water allotment.
- **water application efficiency**
  see **efficiency**.
- **water balance**
  see **hydrologic budget**
- **water budget**
  this is the calculated amount of water to be used for household, based on the type and number of fixtures, landscape requirements, and size of family.
- **water company**
  a private or public entity which provides water , in most cases to properties by means of pipe lines. Some water companies sell water in containers of various sizes.
- **water content of snow**
  see **water equivalent of snow**
- **water conveyance efficiency**
  ratio of the volume of irrigation water delivered by a distribution system to the water introduced into the system.
- **water crop.**
  see **water yield**
- **water cycle**
  the circuit of water movement from the oceans to the atmosphere and to the Earth and return to the atmosphere through various stages or processes such as precipitation, interception, runoff, infiltration, percolation, storage, evaporation, and transportation.
- **water delivery system**
  reservoirs, canals, ditches, pumps., and other facilities to move water.
- **water demand**
  water requirements for a particular purpose, as for irrigation, power, municipal supply, plant transpiration or storage.
- **water equivalent of snow**
  amount of water that would be obtained if the snow should be completely melted. Water content may be merely the amount of liquid water in the snow at the time of observation.
- **water flow**
  the rate of flow of water measured in volume and time.

- **water hammer**
phenomenon which occurs when the velocity of water flowing in pipelines is rapidly changed, usually by a rapid or sudden gate or valve closure, starting or stopping of a pump, or sudden release of air. The resulting pressure waves pass through the water at high velocities and can produce very high momentary positive and/or negative pressures. Shock wave created when the flow of water in a piping system suddenly stops (or changes speed). Usually the result of a fast-closing (or opening) valve.

- **water harvesting**
any of several techniques for increasing the amount of runoff derived from a land area.

- **water holding capacity**
amount of soil water available to plants. See **available soil water**.

- **water holding capacity**
total amount of water held in the soil per increment of depth. It is the amount of water held between field capacity and oven dry moisture level.

- **water horsepower (water power)**
see **horsepower**.

- **water level**
the water surface elevation of a particular water body.

- **water loss**
the difference between the average precipitation over a drainage basin and the water yield from the basin for a given period. The basic concept is that water loss is equal to evapotranspiration, that is, water that returns to the atmosphere and thus is no longer available for use. However, the term is also applied to differences between measured inflow and outflow even where part of the difference may be seen.

- **water meter**
a device used to measure the quantity of water that flows through a pipe.

- **water pollution**
degradation of a body of water by a substance or condition to such a degree that the water fails to meet specified standards or cannot be used for a specific purpose.

- **water pressure**
  see **pressure.**

- **water quality**
  a term used to describe the chemical, physical, and biological characteristics of water, usually in respect to its suitability for a particular purpose.

- **water quality criteria**
  scientifically derived ambient limits for specific pollutants of concern. Criteria are recommended concentrations levels that should not be exceeded in a water body in order to protect aquatic life or human health.

- **water quality-based toxics control**
  an integrated strategy permitting to assess and control the discharge of toxic pollutants to surface waters. There are two approaches:
  1. the whole-effluent approach involves the use of toxicity tests to measure discharge toxicity and
  2. the chemical specific approach involves the use of water quality criteria to limit specific toxic pollutants directly.

- **water requirement**
  the quantity of water, regardless of its source, required by a crop in a given period of time, for its normal growth under field conditions. It includes surface evaporation and other economically unavoidable wastes.

- **water right**
  a legally protected right, granted by law, to take possession of water occurring in a water supply and to divert the water and put it to beneficial use.

- **water spreading**
  the retention of water behind dams or in basins; maintenance of flow in ditches or stream channels, or infusion of water into wells and shafts to develop influent.

- **water storage efficiency**
  see **efficiency.**

- **water supply**
  any quantity of available water.

- **water supply plans**
  these plans are regional water resource and demand analyses. They are District generated, and provide a

detailed evaluation of available water supply and projected demands.

- **water table**

  the level at which the water stays. The very top of the zone of saturation. Water can also be found a few centimetres above this level, due to capillary action. In the presence of a pumping well, the water table drops near the well. This situation is called Drawdown. Under some conditions, a perched water table may exist. This occurs when another confining layer above the 'main' one interrupts the water percolation.

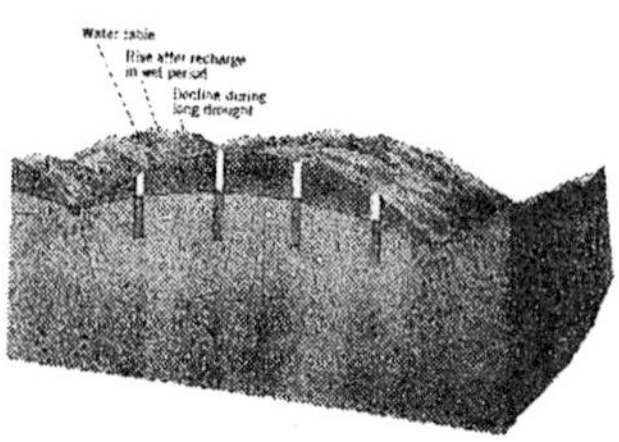

- **water table aquifer**

  an aquifer confined only by atmospheric pressure.

- **water transfers**

  selling or exchanging water or water rights among individuals or agencies.

- **water use**

  water that is used for a specific purpose, such as for domestic use, irrigation, or industrial processing. Water use pertains to human's interaction and influence on the hydrologic cycle, and includes elements such as water withdrawal from surface and ground water sources, water delivery to homes and businesses, consumptive use of water, water released from wastewater-treatment plants, water returned to the environment, and in stream uses, such as using water to produce hydroelectric power.

- **water use efficiency**

  see **efficiency**.

- **water well**

  any artificial excavation constructed for the purpose of exploring for or producing ground water.

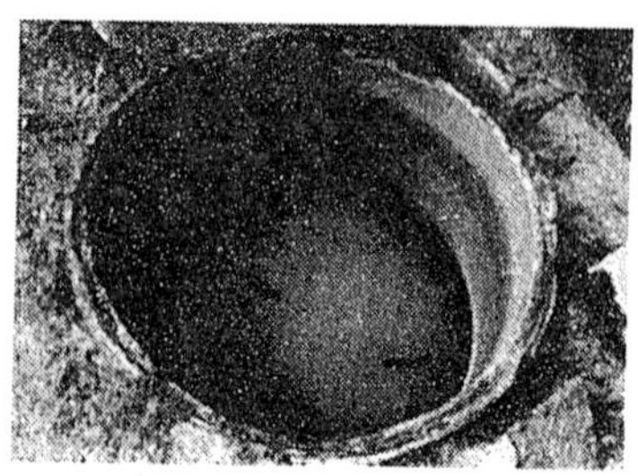

- **water well report**

a report which a water well contractor or landowner who is constructing his own well, submits to a water resources department. It includes the location and dimensions of the well, its flow, a record of geologic materials encountered in drilling, the temperature of the ground water, and other data.

- **water wheel**

a device such as an turbine or similar engine to transform the energy of flowing water into mechanical power.

- **water window**

time of the day available for irrigation to occur.

- **water year**

in Geological Survey reports dealing with surface-water supply, the 12-month period, October 1 through September 30. The water year is designated by the calendar year in which it ends and which includes 9 of the 12 months. Thus, the year ended September 30, 1959, is called the '1959 water year.'

- **water yield (water crop or run out).**

the runoff from the drainage basin, including ground-water outflow that appears in the stream plus ground-water outflow that bypasses the gauging station and leaves the basin underground. Water yield is the precipitation minus the evapotranspiration.

- **water zone**

see **hydrozone**.

- **watercourses**

depressions formed by runoff moving over the surface

of the earth; any natural course that carries water.

- **waterfall**
  a sudden, nearly vertical drop in a stream, as it flows over rock.

- **water-holding capacity**
  the ability of the soil to retain or store water.

- **waterlogging**
  saturation of soil with irrigation water so that the water table rises close to the surface.

- **watershed**
  the land area that drains water to a particular stream, river, or lake. It is a land feature that can be identified by tracing a line along the highest elevations between two areas on a map, often a ridge. Large watersheds, like the Ganges basin contain thousands of smaller watersheds.

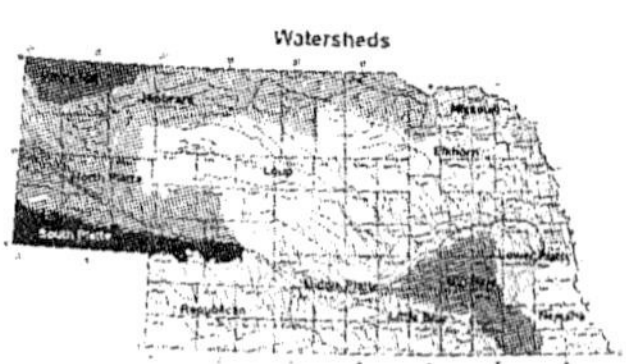

- **Watt-hour (Wh)**
  an electrical energy unit of measure equal to one Watt of power supplied to, or taken from, an electrical circuit steadily for one hour.

- **weather**
  day to day variation in atmospheric conditions.

- **weathering**
  the breakdown of rocks into smaller particles or new chemical substances as a result of exposure to water and air at the earth's surface.

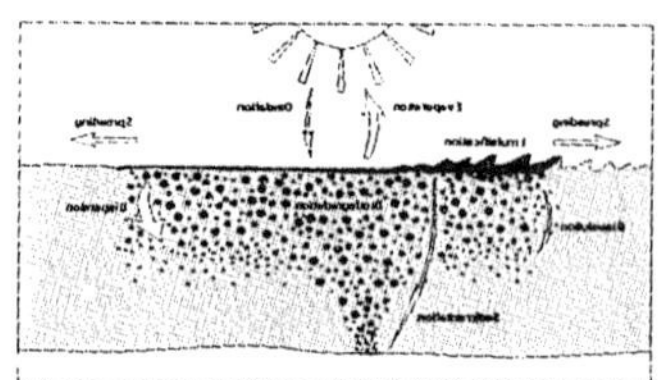

- **weir**
  flow measuring device for open-channel flow. Weirs can be either sharp-crested or broad-crested. Flow opening may be rectangular, triangular, trapezoidal (cipolletti), or specially shaped to make the discharge linear with flow

depth (sutro weir). Calibration is based on laboratory ratings.

- **well**
  a ground-water source, usually a hole dug in the ground down to the water table. Some 'wells' such as artesian wells are probably better classified as 'spri-ngs'.

- **well (water)**
  an artificial excavation put down by any method for the purposes of withdrawing water from the underground aquifers. A bored, drilled, or driven shaft, or a dug hole whose depth is greater than the largest surface dimension and whose purpose is to reach underground water supplies or oil, or to store or bury fluids below ground.

- **well seal**
  top plate of well casing that provides a sanitary seal and support for the drop pipe and pump.

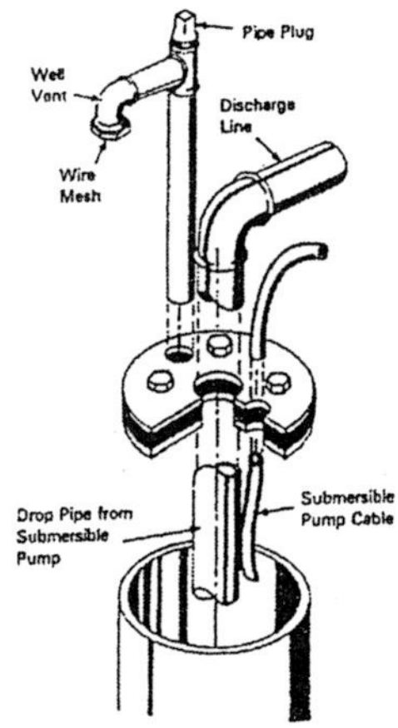

- **wellhead**
  top of the well, at ground level.

- **wellhead protection area**
  a designated area around a public water supply well(s) that is to be protected from contaminants that may adversely affect human health.

- **wet weight (ww)**
weight of soil sample and included soil moisture.

- **wetland**
area that is regularly wet or flooded and has a water table that stands at or above the land surface for at least part of the year, such as a bog, pond, fen, estuary, or marsh.

- **wetlands**
areas that are inundated or saturated by surface or ground water at a frequency and duration sufficient to support a prevalence of vegetation typically adapted for life in saturated soil conditions.

- **wetted area (WA)**
surface area wetted at completion of irrigation.

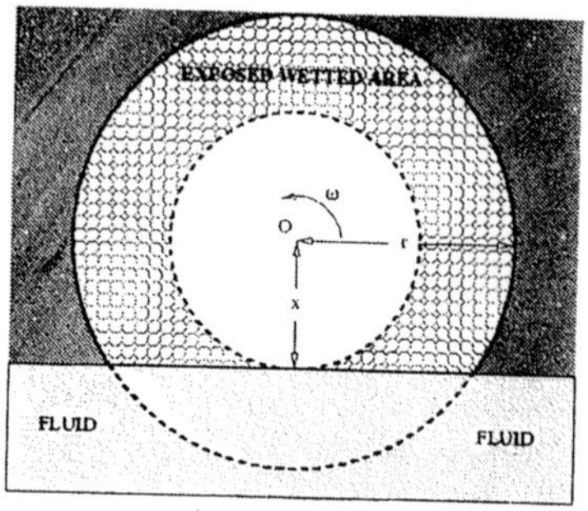

- **wetted diameter**
preferred term diameter of throw.

- **wetted perimeter**
length of the wetted contact between a conveyed liquid and the open channel or closed conduit conveying it, measured in a plane at right angles to the direction of flow.

- **wetting agent**
chemical used to reduce the surface tension of a liquid causing it to make better contact with the desired target.

- **whole-effluent toxicity**
  the aggregate toxic effect of an effluent measured directly by a toxicity test.

- **willingness to pay**
  a method of determining the proxy value of a resource by asking how much users of that resource would be willing to pay to use or not to use it.

- **wilting point**
  1. the soil water content below which plants growing in that soil will remain wilted even when transpiration is nearly eliminated.
  2. may be referred to as the permanent wilting point, describes the amount of water held in the soil which plants are unable to remove. The soil will still contain water, but it is inaccessible to plants.

- **windbreak**
  a line of trees or shrubs planted perpendicular to the prevailing winds, designed to reduce wind velocities and thus reduce wind erosion.

- **winter interest**
  plants that provide texture, colour or other visual effects through leaves, fruit, bark characteristics and other means that enliven the winter landscape.

- **winterisation**
  process of removing water from the irrigation system before the onset of freezing temperatures.

- **wire gauge**
  standard units of measure for wire size. The larger the gauge number, the smaller the wire.

- **withdrawal**
  the removal of water from surface water or groundwater.

- **withdrawal use of water**
  the water removed from the ground or diverted from a stream or lake for use.

- **woody deciduous**
  see **deciduous**.

- **work**

  work done by a force on a particle is defined as the product of the magnitude of the force and the distance through which the particle moves. In hydraulic systems, it can be calculated as the product of the pressure and flow rate.

  brake: measure of work input into a pump or other device

  water: measure of work output from a pump.

- **working pressure**

  pressure of the irrigation system (pis) during operating.

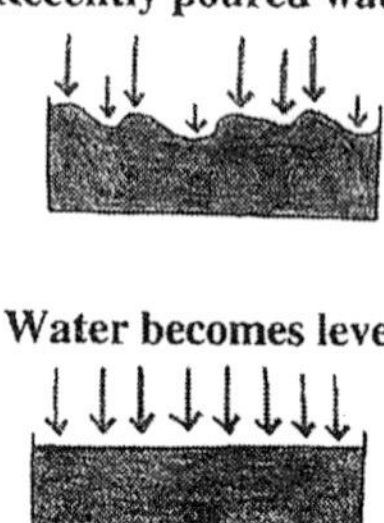

- **working storage (WS)**

  amount of water available in the soil profile for plant use after consideration of MAD. Preferred term is Allowable Depletion [AD]

- **xeriscape**

  creative landscaping for water and energy efficiency and lower maintenance. The seven xeriscape principles are: good planning and design, practical lawn areas, efficient irrigation, soil improvement, use of mulches, low water demand plants, good maintenance.

- **xeriscaping**

  a method of landscaping that uses plants that are well adapted to the local area and are drought-resistant. Xeriscaping is becoming more popular as a way of saving water at home.

- **year**

  see **climatic year; water year**

- **yield**

  the quantity of water expressed either as a continuous rate of flow (cubic feet per second, etc.) or as a volume per unit of time. It can be collected for a given use, or uses, from surface or groundwater sources on a watershed.

- **zone**

  section of an irrigation system served by a single control valve. Zones are comprised of similar sprinkler types and plant material types with similar water requirements and types.

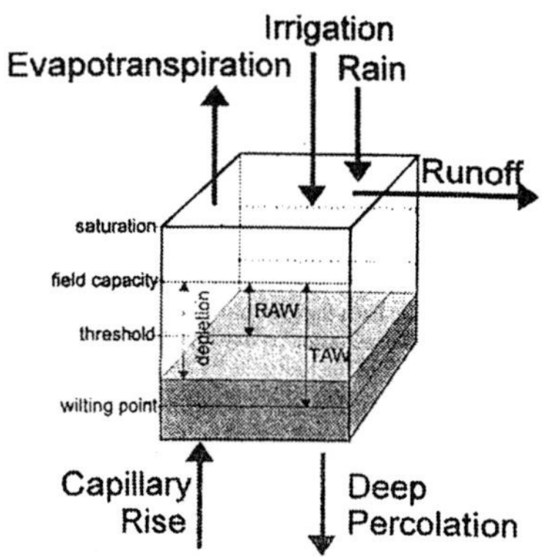

- **zone of aeration**

  a region in the Earth above the water table. Water in the zone of aeration is under atmospheric pressure and will not flow into a well.

- **zone of saturation**

  the zone in which the functional permeable rocks are saturated with water under hydrostatic pressure. Water in the zone of saturation will flow into a well, and is called ground water.

- **zoning**

  a system of land-use management in which land is classified according to permitted uses.

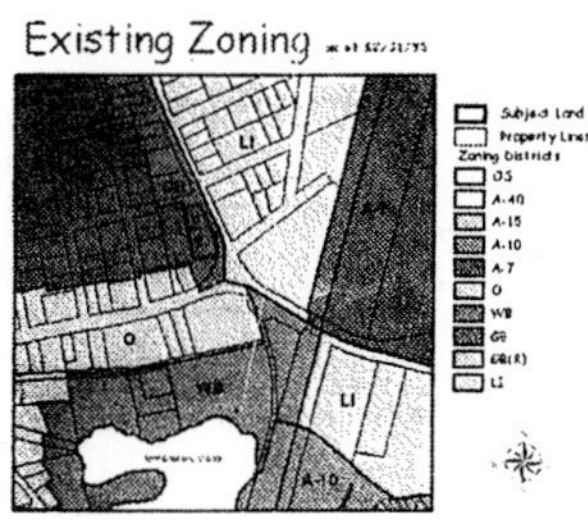

Notes